ALGEBRA
Practice Exercises

Thomas E. Campbell

J. WESTON
WALCH
PUBLISHER
Portland, Maine

To Katie, Beth, and Meg

REPRODUCTION OF BLACKLINE MASTERS

These blackline masters are designed for individual student use and intended for reproduction by dry copier, liquid duplicating machine, or other means.

J. Weston Walch, Publisher, therefore transfers limited reproduction rights to the purchaser of these masters at the time of sale. These rights are granted only to a single classroom teacher, whether that teacher is the purchaser or the employee of a purchasing agent such as a school district. The masters may be reproduced in quantities sufficient for individual use by students under that teacher's direct classroom supervision.

Reproduction of these masters for use by other teachers or students is expressly prohibited and constitutes a violation of United States Copyright Law.

---

1 2 3 4 5 6 7 8 9 10

ISBN 0-8251-2850-1

Copyright © 1996
J. Weston Walch, Publisher
P.O. Box 658 • Portland, Maine 04104-0658

Printed in the United States of America

Contents

ALGEBRA .... *iii*
Foreword.. *vii*
Answer Key .... *ix*

1. Identifying Algebra Vocabulary and Properties 1
2. Fractions, Decimals, and Percents 2
3. Solving Percent Problems for the Part 3
4. Solving Percent Problems for the Rate 4
5. Solving Percent Problems for the Whole 5
6. Solving Percent Problems 6
7. Order of Operations 7
8. Evaluation of Expressions 8
9. Solving Simple Equations 9
10. Solving Absolute Value Equations 10
11. Exponents 11
12. Operations with Negative Numbers 12
13. Distribution and Double Distribution 13
14. Combining Like Terms 14
15. Divisibility Rules 15
16. Factoring to Primes 16
17. Finding the Least Common Multiples 17
18. Finding Common Factors 18
19. Factoring Out the Greatest Common Factor 19
20. Multiplying Binomials 20
21. Factoring by Association 21
22. Factoring Trinomials by Undoing Distribution 22
23. Solving Equations by Factoring and the Difference of Two Squares 23
24. Solving Quadratic Equations by Factoring 24
25. A Variety of Factoring Strategies 25
26. Solving Complex Equations 26
27. Simplifying Radicals 27
28. The Pythagorean Theorem 28
29. Solving Quadratic Equations by Completing the Square 29
30. Solving Quadratic Equations by the Quadratic Formula 30
31. Solving Quadratic Equations 31
32. The Cartesian Coordinate System 32

33. The Distance Formula 33
34. Synthetic Division 34
35. Two Variable Equations and the Chart Graphing Method 35
36. The Slope of a Graph 36
37. The Slope/Y-Intercept Method 37
38. Finding the Equation from the Graph 38
39. Parallels and Perpendiculars 40
40. Finding Intersections by Plotting the Graphs 43
41. Solving Simultaneous Equations by the Substitution Method 46
42. Solving Simultaneous Equations by the Elimination Method 47
43. Solving Word Problems 48
44. Scientific Notation 49
45. Simplifying Rational Expressions 50
46. Solving Simple Inequalities 51
47. Solving Compound Inequalities 52
48. Solving Compound Absolute Value Inequalities 53
49. Graphing Linear Inequalities 54
50. Graphing Systems of Linear Inequalities 57

Foreword

All too often, I find myself making up problems for further work because a primary textbook just can't predict how much practice this year's set of students will need to master a particular concept. Over the years, I have built a resource for these occasions. The problems in this book are designed to give students practice in the basics of algebra, *and* to get them thinking a bit about what lies behind a particular skill or idea. Students can work on the exercises individually or in small groups. They are also sometimes asked to write explanations, not just results of calculations. I hope you will find these worksheets useful as a resource.

—T.E.C.
Portland, ME 1995

Answer Key

1. Identifying Algebra Vocabulary and Properties

1. (a) Answers will vary. 4, *a, x, x²*, 3, *c, y*
   (b) Answers will vary. Addition and multiplication.
   (c) *x*
   (d) Answers will vary. Usually 3 or 4.
2. (a) – (ii)
   (b) – (iv)
   (c) – (v)
   (d) – (vi)
   (e) – (i)
   (f) – (iii)
3. 97 × 12 + 3 × 12 = (97 + 3) × 12 = 100 × 12 Distributive property.
4. 0
5. Answers will vary. Addition and multiplication.
6. Answers will vary. Subtraction and division.
7. Answers will vary. Addition and multiplication.
8. Answers will vary. Subtraction and division.
9. No. 3 and 2 are in Z^+ and $3/2$ is not.

2. Fractions, Decimals, and Percents

1. .6
2. 86%
3. 303/500
4. 25%
5. 9/20
6. .575
7. 43/50
8. .8333333→
9. 3/4
10. .0007
11. 20%
12. 37.5%
13. 2/3
14. .142857142857→
15. 125%
16. 3/11
17. 44.4444→%
18. .272727→
19. 31.25%
20. 35/99

3. Solving Percent Problems for the Part

1. 72
2. 72
3. 44
4. 17
5. 72
6. 319
7. 33
8. 15
9. 90
10. 49
11. 242
12. 243
13. 168 marbles
14. About 30.5 points were lost on the Dow Jones.
15. 75,573 ninth graders require vision correction.

4. Solving Percent Problems for the Rate

1. $K = .8$
2. $A = .25$
3. $T = .12$
4. $I = 25$
5. $E = 16\frac{2}{3}$
6. $B = 6\frac{1}{4}$
7. $E = 33\frac{1}{3}$
8. $T = 25$
9. $H = 20$
10. 4%
11. 87.5%
12. $6\frac{2}{3}$%
13. Clinton, 43.0% Bush, 37.4%   Perot, 19.6%
14. The farmer receives about 21% of the supermarket price.
15. About 12.7% of the graduates were math majors.

5. Solving Percent Problems for the Whole

1. $X = 24$
2. $Z = 130$
3. $P = 500$
4. $Y = 220$
5. $B = 80$
6. $K = 1375$
7. $M = 84$
8. $J = 33\frac{1}{3}$
9. $H = 66\frac{2}{3}$
10. 500
11. 150
12. 350
13. About 322 pregnancies in Vermont result in multiple births.
14. Tyler has 80 postcards.
15. There are about 233,333 14-year-olds in the United States.

6. Solving Percent Problems

1. $D = 30$
2. $M = .26$
3. $Y = 108$
4. $Y = 198$
5. $K = 46.875$
6. $J = 44$
7. $N = 70$
8. $J = 66\frac{2}{3}$
9. $L = 98$
10. 35 is 140% of 25.
11. 243 is 108% of 225.
12. 260 is 500% of 52.
13. You can expect to roll a four $8\frac{1}{3}$% of the time.
14. About 62,400 Jaguar customers are satisfied.
15. There were about 5,609 fans at the game.
16. The meal cost $52.07.

x   Algebra Practice Exercises

7. Order of Operations

1. 48
2. 14
3. 21
4. 7
5. 64
6. 1
7. −3
8. 16
9. 43
10. Answers will vary. An example is 77 ÷ 7 + 7.

Answers for 11–17 will vary. One example for each is:

11. 32 − 13 + (2) 1 = 21
12. 3 (2 − 1) 3 + 21 = 30
13. (32 − 1) 3 + 21 = 114
14. 32 − 13 + 21 = 40 (No need to insert parentheses)
15. (3 (2 − 13) + 2) 1 = −31
16. 3 (2) − 13 + 2 (1) = −5
17. 3 (2 − 1) (3 + 2) 1 = 15
18. Answers will vary. Students should be encouraged to check each other's work.

8. Evaluation of Expressions

1. 8
2. 19
3. −20
4. 3
5. −31
6. −95
7. 4
8. −20
9. 475
10. −72
11. −10
12. −155
13. 45
14. 56
15. 68
16. −52
17. −391
18. −70

9. Solving Simple Equations

1. $X = 4$
2. $A = 5$
3. $Y = 20$
4. $M = 7$
5. $E = 7$
6. $G = 3$
7. $A = 7$
8. $N = 3$
9. $B = 11$
10. $U = 10$
11. $T = 11$
12. $L = 7$
13. $E = -13$
14. $R = 10$
15. $I = 11\frac{5}{8}$
16. $L = -5\frac{4}{7}$
17. $Y = 2\frac{19}{24}$

10. Solving Absolute Value Equations

1. $X = -8$ or 16

   <-- | -- | -- | -- | -- | -- | -- | -- | -- | -- | -- | -- | -- | -- | -- | -- | -->
   -12 -10 -8 -6 -4 -2 0 2 4 6 8 10 12 14 16 18 20

2. $B = -11$ or 1

   <-- | -- | -- | -- | -- | -- | -- | -- | -- | -- | -- | -- | -- | -- | -- | -- | -->
   -14 -13 -12 -11 -10 -9 -8 -7 -6 -5 -4 -3 -2 -1 0 1 2

3. $E = -16$ or 28

   <-- | -- | -- | -- | -- | -- | -- | -- | -- | -- | -- | -- | -- | -- | -- | -- | -->
   -20 -16 -12 -8 -4 0 4 8 12 16 20 24 28 32 36 40 44

4. $T = 11$ or 21

   <-- | -- | -- | -- | -- | -- | -- | -- | -- | -- | -- | -- | -- | -- | -- | -- | -->
   90 10 11 12 13 14 15 16 17 18 19 20 21 22 23 24 25

5. $H = -4$ or 10

   <-- | -- | -- | -- | -- | -- | -- | -- | -- | -- | -- | -- | -- | -- | -- | -- | -->
   -6 -4 -2 0 2 4 6 8 10 12 14 16 18 20 22 24 26

6. $A = -12$ or 15

   <-- | -- | -- | -- | -- | -- | -- | -- | -- | -- | -- | -- | -- | -- | -- | -- | -->
   -18 -15 -12 -9 -6 -3 0 3 6 9 12 15 18 21 24 27 30

7. $N = 1.6$ only

   <-- | -- | -- | -- | -- | -- | -- | -- | -- | -- | -- | -- | -- | -- | -- | -- | -->
   1.0 1.1 1.2 1.3 1.4 1.5 1.6 1.7 1.8 1.9 2.0 2.1 2.2 2.3 2.4 2.5 2.6

8. $Y = .5$ only

   <-- | -- | -- | -- | -- | -- | -- | -- | -- | -- | -- | -- | -- | -- | -- | -- | -->
   0 .1 .2 .3 .4 .5 .6 .7 .8 .9 1.0 1.1 1.2 1.3 1.4 1.5 1.6

xii   Algebra Practice Exercises

9. $I = -7$ or 1

```
       ●                               ●
<--|--|--|--|--|--|--|--|--|--|--|--|--|--|--|--|-->
-10 -9 -8 -7 -6 -5 -4 -3 -2 -1  0  1  2  3  4  5  6
```

10. $V = -2\frac{1}{2}$ or 5

```
       ●                     ●
<--|--|--|--|--|--|--|--|--|--|--|--|--|--|--|--|-->
 -4 -3 -2 -1  0  1  2  3  4  5  6  7  8  9 10 11 12
```

11. $E =$ No solution

12. $S = -1.75$ or 7.75

```
          ●                         ●
<--|--|--|--|--|--|--|--|--|--|--|--|--|--|--|--|-->
 -3 -2 -1  0  1  2  3  4  5  6  7  8  9 10 11 12 13
```

13. $I = \frac{1}{4}$ or $3\frac{3}{4}$

```
          ●                                          ●
<--|--|--|--|--|--|--|--|--|--|--|--|--|--|--|--|-->
-¼  0  ¼  ½  ¾  1 1¼ 1½ 1¾ 2 2¼ 2½ 2¾ 3 3¼ 3½ 3¾
```

14. $L = \frac{1}{3}$ or 9

```
          ●                     ●
<--|--|--|--|--|--|--|--|--|--|--|--|--|--|--|--|-->
 -1  0  1  2  3  4  5  6  7  8  9 10 11 12 13 14 15
```

15. $Y = 3\frac{2}{3}$ or 5

```
                        ● ●
<--|--|--|--|--|--|--|--|--|--|--|--|--|--|--|--|-->
 -5 -4 -3 -2 -1  0  1  2  3  4  5  6  7  8  9 10 11
```

16. $T = -17$ or 15

```
          ●                     ●
<--|--|--|--|--|--|--|--|--|--|--|--|--|--|--|--|-->
-25 -21 -17 -13 -9 -5 -1  3  7 11 15 19 23 27 31 35 39
```

17. $C = -6$ or 12

```
          ●                           ●
<--|--|--|--|--|--|--|--|--|--|--|--|--|--|--|--|-->
-10 -8 -6 -4 -2  0  2  4  6  8 10 12 14 16 18 20 22
```

11. Exponents

1. $12A^6B^8$
2. $-10X^6Y^4Z^7$
3. $125X^9Y^{18}$
4. $256F^{16}G^{20}H^{12}$
5. $\dfrac{2}{3U^2D^7}$
6. $\dfrac{5F^5}{3G^3R}$
7. $\dfrac{4M^5}{P^5}$
8. $\dfrac{X^2}{55Z^3}$
9. Answers will vary. Mention should be made of the associative property.
10. Answers will vary. Mention should be made of the fact that $W^xW^x = W^{2x}$, etc.
11. $\dfrac{1 + C^4D^7}{C^8D^3}$
12. $4G^4H^4 - H^6G^5$
13. 4
14. 27
15. 5^2
16. 7^{2e}

12. Operations with Negative Numbers

1. 5
2. 10
3. −11
4. −12
5. 19
6. 6
7. 18
8. 36
9. 76
10. −49
11. −2
12. −24
13. 0
14. 1
15. 24
16. −11
17. 66
18. −4 (32−18)
19. Answers will vary.

xiv   Algebra Practice Exercises

13. Distribution and Double Distribution

1. $12X + 20Y$
2. $12BX - 21B$
3. $12C + 18CD$
4. $6C^2R - 9CR^2$
5. $15X^3 - 10X^2 - 5X$
6. $21Y^2Z^2 - 35Y^4Z$
7. $12D^2 - 28DE + 8D^3$
8. $35N - 15N^2 - 10N^3$
9. $X^2 - X - 6$
10. $Z^2 - 11Z + 28$
11. $N^2 + 13N + 12$
12. $K^2 + 2K - 24$
13. $2X^2 + X - 15$
14. $4NM - 3M + 4N - 3$
15. $10Q^2 - 32Q + 6$
16. $2W^2 - 3VW + 13W - 21V - 7$
17. $2CF + 2C - 3DF - 3D + 2F + 2$
18. Answers will vary, but some mention should be made of "like terms" and having two shared terms in the problem.

14. Combining Like Terms

1. $X^2 - 5X + 5$
2. $K^2 - 4K - 21$
3. $17X - 3Z$
4. $6Z + 4Z^2$
5. $2XY - 2ZY + 2XZ$
6. $7AB - 9B^2 - 7A$
7. $-2X^2 - 2Y^2 + 3X - 2Y + 7 - 2XY$
8. $X + 5 - 3XY$
9. $8VD - 6V^2D - 3$
10. Cannot be simplified.
11. Cannot be simplified.
12. $7T^2H + 30TH - 6TH^2$
13. $6X^2 + 3XY - 5Y^2$
14. $6A^2 - 5AB - 6B^2$
15. $84V^2W - 30VW^2 + 6VW$
16. $-43GH - 46H^2 + 3G^2$
17. $2C^2 - 13DC + 4C + 15D^2 - 13D + 2$
18. $6W^2Y^2 - 9W^2Y + 23WY - 14WY^2 + 14Y - 12W - 8$

15. Divisibility Rules

1. No. The sum of 3, 4, and 9 is 16, which is not divisible by 3.
2. Yes. 8 is even.
3. Yes. 376 − 343 = 33, which is divisible by 11.
4. No. 578 is even, but 5 + 7 + 8 = 20, which is not divisible by 3.
5. Yes. 6,525 ends in a 5.
6. Yes. 7,865,900 ends in a 0.
7. No. 300 − 492 = −192, which is not divisible by 7.
8. Yes.
9. No.
10. No.
11. Yes. 765 − 934 + 435 = 266, which is divisible by 7.
12. No.
13. Yes.
14. Yes.
15. No.
16. Yes.
17. No, because it is not divisible by 3.

16. Factoring to Primes

1. $2^3 \times 3^3$
2. $2^3 \times 3^2 \times 5 \times 7$
3. $3^2 \times 5^4$
4. $2 \times 3^2 \times 17$
5. 2^8
6. $2 \times 3^4 \times 7$
7. $2 \times 3 \times 5^2 \times 17$
8. $2 \times 7^2 \times 11$
9. 11×13^2
10. $2^4 \times 3^2 \times 7 \times 11$
11. Prime
12. 17×89
13. 7×113
14. 3×263
15. $2^8 \times 3 \times 5$
16. $3^5 \times 7^2$
17. $3^2 \times 7 \times 13^2$
18. $3^3 \times 5^2 \times 7^2$
19. $5^2 \times 13 \times 17^2$
20. 2^{21}

17. Finding the Least Common Multiples

1. 102
2. 72
3. 84
4. 72
5. 1,008
6. 42
7. 2,304
8. 7,425
9. 896
10. 676
11. 588
12. 882
13. 3,328
14. 61,440
15. 108
16. 2,730
17. 2,604
18. 6,480
19. Answers will vary: 9, 45, 99, 495
20. Answers will vary: 99, 198, 693, 1,386

18. Finding Common Factors

1. $2X$
2. $7BC$
3. DE
4. $9XY^2Z$
5. $6EF^2$
6. 3
7. $25XM$
8. 1 (No common factors)
9. $9MP$
10. $13UVM$
11. $7T^2C(4EC^2 - 5H^4)$
12. $2XY(18X - 7Y^2 - 12)$
13. $5XD^2F(15X^3 + 45XD^3F^3 - 16DF)$
14. $21J^3K^2L(21JK^3 - 3 + 5K^5L^2)$
15. $64G^4H^5(4G^3 + P^2G^2 - 16H^2D^{10})$
16. $G(504R^4GP - 18P^2H^5 + 77GHR^6)$
17. $15SCT^2(24S^4 + 9TS - 4C^5T^3)$

19. Factoring Out the Greatest Common Factor

1. $X^2(2-3X)$
2. $5(3a^2+b^2)$
3. $3W^2(4+W)$
4. $7m^2n^2(n-3m)$
5. $t^4s^4(t^2s+1)$
6. $Q(121P^3Q+169D^3)$
7. $XY^2Z(X^2Y-Z)$
8. $11a^3x^2(a-8x)$
9. $36x^2y(4y^4-x)$
10. $13d^2h^2(2-3h)$
11. $7m^2n^2(n-3m+9)$
12. $11Q(11P^3Q+15D^3+7X^3Q^2)$
13. Not simplifiable. No factor is shared by all three terms.
14. $XYZ(3X^2Y^2Z^2-2XYZ-1)$
15. $abc(a^2b-ac^2+b^2c)$
16. Simplifies to $18X^5-9Y^3$, and then factors as $9(2X^5-Y^3)$
17. $2β^2©^2(8β+9β©^2+4©^2)$
18. $50X^2Y(33Y^2+5X-17)$
19. $1024X(1024X^2+Y^2-32XZ)$
20. $X^4Y^3-X^2Y^4+X^2Y^3$ or $X^3Y^3-X^2Y^4+X^2Y^3$

20. Multiplying Binomials

1. X^2-6X-7
2. $W^2+11W+30$
3. $K^2-13K+40$
4. $A^2+4A-12$
5. T^2-9
6. $H^2-10H+25$
7. $E^2+3E-108$
8. $R^2+4R-77$
9. J^2-64
10. N^2-N-30
11. $E^2-8E+16$
12. $2C^2-C-3$
13. $3H^2-8H+4$
14. $6A^2+5A-6$
15. $12S^2-31S+20$
16. $3E^2-26E-9$
17. $25+20J-12J^2$
18. $9L^2-49$
19. $25Y^2+30Y+9$
20. $81T^2+9T-20$

21. Factoring by Association

1. $(X + 2)(Y + 3)$
2. $(A + 4)(B - 6)$
3. $(2M + 3)(X - 5)$
4. $(4q - 1)(t + 6)$
5. $(p - 3)(r + 4)$
6. $(3L - 4)(X + 7)$
7. $(V + 3)(2W - 6)$
8. $(3K^2 - 7)(2K + 1)$
9. $2(6Z^2 - 7)(Z + 2)$
10. $(a + b + c)(c - 6)$
11. $(m + n - 3)(2n - m)$
12. $(2XY - Z)(2Z + W)$
13. $12XK - 28X + 9K - 21 = (3K - 7)(4X + 3)$
14. $2XW - WY + 3Y - 6X = (2X - Y)(W - 3)$
15. $8Z^2 - 6CZ - 2Z + 16Z - 12C - 4 = (4Z - 3C - 1)(2Z + 4)$

22. Factoring Trinomials by Undoing Distribution

1. $(X + 4)(X + 5)$
2. $(Z + 4)(Z + 8)$
3. $(A + 9)(A + 9)$ or $(A + 9)^2$
4. $(M - 9)(M - 9)$ or $(M - 9)^2$
5. $(Q + 4)(Q + 14)$
6. $(L - 7)(L - 8)$
7. $(N - 16)(N + 1)$
8. $(C + 8)(C - 2)$
9. $(T + 3)(T - 9)$
10. $(Y + 3)(Y + 9)$
11. $(L + 4)(L - 16)$
12. $(E + 32)(E - 2)$
13. $(R + 27)(R + 3)$
14. No solution: No two positive whole numbers have product = 1 and sum = 1.
15. $(A - 1)(A - 1)$ or $(A - 1)^2$
16. $(N + 2)(N + 1)$
17. $(\spadesuit - 10)(\spadesuit - 3)$
18. $G^2 - 10G + 21 = (G - 7)(G - 3)$
19. $H^2 + 12H + 27 = (H + 3)(H + 9)$
20. $K^2 - 10K + 21 = (K - 7)(K - 3)$

23. Solving Equations by Factoring and the Difference of Two Squares

1. $(X + 20)(X - 2)$
2. $(a - 7)(a - 5)$
3. $(y - x)(y - x)$ or $(y - x)^2$
4. $(y - 4)(y + 4)$
5. $(z - 11)(z + 11)$
6. $(H - 7)(H + 7)$
7. $(F + 24)(F - 24)$
8. Answers will vary. This isn't a *difference* of two squares, it's a *sum*-of-two-squares problem.
9. $(z^2 + 4)(z + 2)(z - 2)$
10. $(G^2 + 9)(G + 3)(G - 3)$
11. $m = 5$ or -2
12. $z = 5$ or 6
13. $e = 5$ only
14. $e = 5$ or -5
15. $e = 5$ or -5
16. Answers will vary. Mention should be made of what 0e means and how including 0e allows the pattern that has been used to solve these to be continued.
17. $M = \pm 2\sqrt{2}$

24. Solving Quadratic Equations by Factoring

1. $X = -2$ or $X = 7$
2. $X = -2$ or $X = -5$
3. $X = 6$ or $X = -3$
4. $X = -6$ or $X = 2$
5. $X = 3$ or $X = 9$
6. $X = 3$ or $X = 4$
7. $X = -4$
8. $X = 6$ or $X = -6$
9. $X = -2$ or $X = -12$
10. $X = 9$
11. $X = 4$ or $X = 14$
12. $X = 20$ or $X = -2$
13. $X = 6$ or $X = 12$
14. $X = -1$ or $X = -17$
15. $X = -8$ or $X = -10$
16. $X = 7$ or $X = 11$

25. A Variety of Factoring Strategies

1. $(X-5)(X+5)$
2. $(2W-9)(2W+9)$
3. $(M^2N^2-U)(M^2N^2+U)$
4. Cannot be factored.
5. $(P^2-6)(P^2+6)$
6. Cannot be factored.
7. $(Z^8+1)(Z^4+1)(Z^2+1)(Z+1)(Z-1)$
8. Number 4 can't be factored because it is the *sum* of two squares, and number 6 can't be factored because 8 is not a perfect square.
9. $(H-2)(H^2+2H+4)$
10. $(5-Y)(25+5Y+Y^2)$
11. $(A^2-3)(A^4+3A^2+9)$
12. $(6-7R)(36+42R+49R^2)$
13. $(X+1)(X^2-X+1)$
14. $(K+4)(K^2-4K+16)$
15. $(M^2+5)(M^4-5M^2+25)$
16. $(11T+7)(121T^2-77T+49)$
17. $(2-A)(B+3)$
18. $(X-3)(X+Y)$
19. $(2e-3)(f+6)$
20. $(X-2+2Y)(X-2-2Y)$

26. Solving Complex Equations

1. $X=3$
2. $Y=2\frac{3}{4}$
3. $G=2\frac{7}{11}$
4. $K=3\frac{4}{5}$
5. $M=3\frac{2}{27}$
6. No solution.
7. $U=-1\frac{10}{19}$
8. $t=5\frac{14}{19}$
9. $G=-2$
10. $C=1\frac{1}{12}$
11. $T=2\frac{87}{88}$
12. $r=1\frac{1}{4}$
13. $J=1\frac{5}{11}$
14. $A=3\frac{7}{12}$
15. $E=\frac{631}{822}$

27. Simplifying Radicals

1. 24
2. 27
3. $9\sqrt{2}$
4. $11\sqrt{5}$
5. $5\sqrt{42}$
6. $16\sqrt{2}$
7. $20\sqrt{3}$
8. $\dfrac{2\sqrt{21}}{7}$
9. $\dfrac{2}{3}$
10. 12.5
11. $2\sqrt{10}$
12. $\sqrt{2}$
13. 10
14. $7\sqrt{5} - 4\sqrt{2}$
15. $27\sqrt{3}$
16. $(-\sqrt{14})/28$
17. $30\sqrt{3} - 45\sqrt{5} - 4\sqrt{30} + 30\sqrt{2}$
18. $72 - 120\sqrt{2} + 18\sqrt{6} - 60\sqrt{3}$
19. $\dfrac{5\sqrt{5} - 5}{4}$
20. $\dfrac{7\sqrt{12} - 21}{3}$

28. The Pythagorean Theorem

1. $W = 10$
2. $W = 13$
3. $W = \sqrt{346}$
4. $W = 15$
5. $W = \sqrt{1145}$
6. $W = \sqrt{932} = 2\sqrt{233}$
7. $W = \sqrt{171}$
8. $W = \sqrt{490} = 7\sqrt{10}$
9. $W = \sqrt{1007}$
10. $W = \sqrt{250} = 5\sqrt{10}$
11. $W = \sqrt{585} = 3\sqrt{65}$
12. $W = \sqrt{483}$

29. Solving Quadratic Equations by Completing the Square

1. $X = 6$ or $X = -2$
2. $X = -6$ or $X = -2$
3. $X = 4$ or $X = 2$
4. $X = -2.5$ or $X = -3.5$
5. $X = 7$ or $X = 5$
6. $X = 6$ or $X = -3$
7. $X = -2$ or $X = -3$
8. $X = 3$ or $X = -12$
9. $X = -2 \pm \sqrt{13}$
10. $X = 3 \pm \sqrt{3}$
11. $X = 4 \pm \sqrt{21}$
12. $X = 3/2 \pm \sqrt{3}/2$
13. $X = -7/2 \pm \sqrt{41}/2$
14. $X = -3/2 \pm \sqrt{5}/2$
15. $X = +3/2 \pm \sqrt{17}/2$
16. $X = 1$ or $X = -3.5$

30. Solving Quadratic Equations by the Quadratic Formula

1. $X = 4$ or $X = -3$
2. $X = -3$ or $X = -5$
3. $X = 1$ or $X = 3$
4. $X = -1.5$ or $X = -4.5$
5. $X = 1.5$
6. $X = 5/2 \pm \sqrt{53}/2$
7. $X = -13/2 \pm \sqrt{145}/2$
8. $X = -7/2 \pm \sqrt{13}/2$
9. $X = 1$ or $X = -7/3$
10. $X = 6/5 \pm \sqrt{11}/5$
11. $X = 7/4 \pm \sqrt{161}/4$
12. $X = 3/2 \pm \sqrt{3}/2$
13. No solution.
14. $X = -3 \pm \sqrt{22}/2$
15. $X = 5/3 \pm \sqrt{31}/3$
16. $X = -7/6 \pm \sqrt{103}/6$

31. Solving Quadratic Equations

1. $X = 4$ or $X = -3$
2. $X = -3$ or $X = -5$
3. $X = 1$ or $X = 3$
4. $X = -1.5$ or $X = -4.5$
5. $X = 1.5$
6. $X = 5/2 \pm \sqrt{53}/2$
7. $X = -13/2 \pm \sqrt{145}/2$
8. $X = -7/2 \pm \sqrt{13}/2$
9. $X = 1$ or $X = -7/3$
10. $X = 6/5 \pm \sqrt{11}/5$
11. $X = 7/4 \pm \sqrt{161}/4$
12. $X = 3/2 \pm \sqrt{3}/2$
13. No solution.
14. $X = -3 \pm \sqrt{22}/2$
15. $X = 5/3 \pm \sqrt{31}/3$
16. $X = -7/6 \pm \sqrt{103}/6$

32. The Cartesian Coordinate System

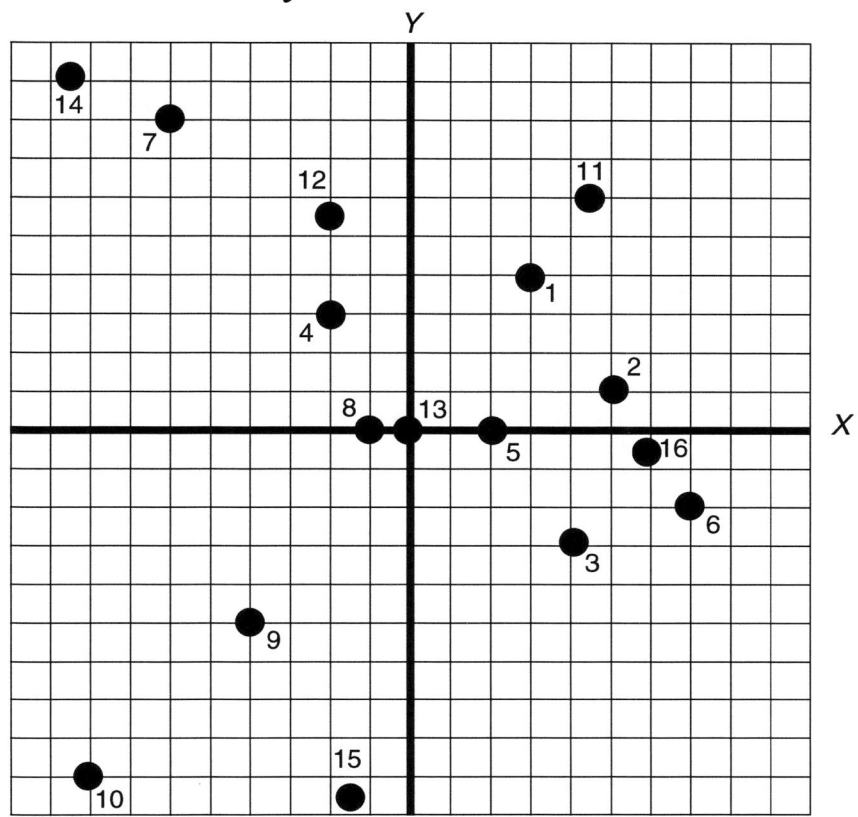

33. The Distance Formula

1. 5 units
2. $\sqrt{13}$ units
3. $\sqrt{58}$ units
4. $\sqrt{73}$ units
5. $5\sqrt{13}$ units
6. 13 units
7. $15\sqrt{2}$ units
8. 5 units
9. $8\sqrt{2}$ units
10. $\sqrt{[(X-3)^2 + (Y-6)^2]}$
11. $\sqrt{[(X-P)^2 + (Y-Q)^2]}$

34. Synthetic Division

1. $X^2 + 2X + 1$ r. 0
2. $3X^2 - 4X + 13$ r. -25
3. $T + 1$ r. 0
4. $7Y^2 + 26$ r. 55
5. $X^3 - X^2 - \frac{7}{3}X - \frac{2}{9}$ r. $-1\frac{4}{9}$
6. $2X^3 + 5X^2 + 20X + 89$ r. 348
7. $X^4 + X^3 + X^2 + X + 1$ r. 0
8. $2X^5 + 9X^4 + 27X^3 + 74X^2 + 222X + 669$ r. 2006
9. $3X^3 - 7.5X^2 + 16.25X - 40.625$ r. 190.125
10. $3X^4 + 8X^3 + 25X^2 + 69X + 219$ r. 665
11. $2X^4 + X^3 + 4X^2 + 2X - 2\frac{1}{2}$ r. $-2\frac{1}{2}$
12. $3X^4 - 10.5X^3 + 23.75X^2 - 19.625X + 52.9375$ r. -246.8125

35. Two Variable Equations and the Chart Graphing Method
(Answers will vary.)

1.

X	Y
2	0
0	-2
3	1

2.

X	Y
3	1
5	4
1	-2

3.

X	Y
0	.5
2	3
6	8

4.

X	Y
2	0
0	-4
1	-2

5.

X	Y
3	3
0	.6
-.75	0

6.

X	Y
8	0
8	-1
8	1

7.

X	Y
-1	1
0	2.2
4	7

8.

X	Y
2	2
1.4	0
-1	-8

9.

X	Y
0	-2
1	-2
-1	-2

10.

X	Y
0	2
6	0
9	-1

36. The Slope of a Graph

1. Slope = 1/3
2. Slope = -4/4 or -1/1
3. Slope = -1/-3 or 1/3
4. Slope = -1/-2 or 1/2
5. Slope = 0/2 or 0 or horizontal
6. Slope = -2/1
7. Slope = 1/4
8. Slope = 2/2 or 1/1
9. Slope = 0/4 or 0 or horizontal
10. Slope = -2/-1 or 2/1
11. Slope = 1/3
12. Slope = 1/0 or undefined or vertical

37. The Slope/Y-Intercept Method

1. Slope = 2/1
2. Slope = -1/2
3. Slope = 3/1
4. Slope = -1/2
5. y – intercept is -3
6. y – intercept is 3
7. y – intercept is 2.5
8. y – intercept is -1.75
9. 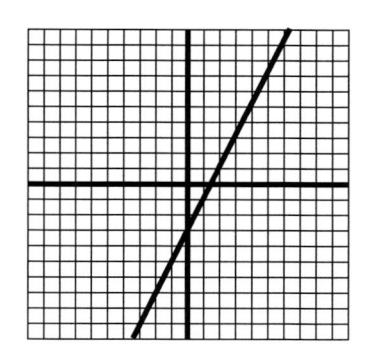
10. 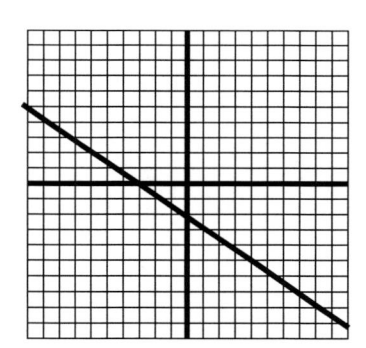
11.

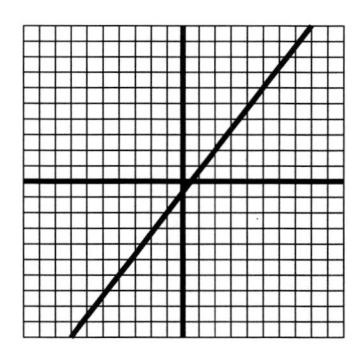

38. Finding the Equation from the Graph

1. $X = Y$
2. $Y = 3/4 X$
3. $Y = -1/4 X + 3/2$
4. $Y = -2/1 X + 12$
5. $Y = -1/2 X + 7/2$
6. $Y = 3/1 X - 15$
7. $Y = -3/7 X$
8. $Y = 3X - 9$
9. $Y = 4/5 X$
10. $Y = 4$

39. Parallels and Perpendiculars

1. $Y = -2X + 7$ 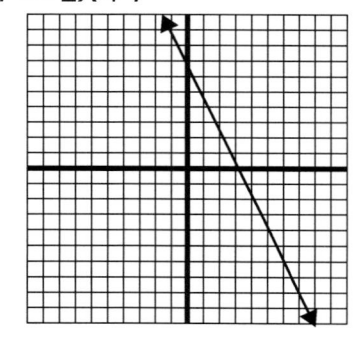
2. $Y = -3X + 8$ 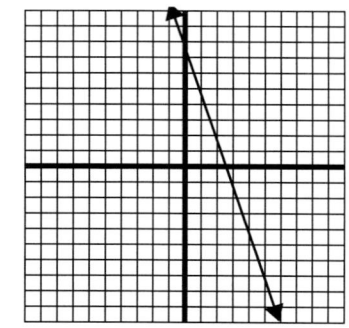
3. $Y = 5/3 X$
4. $Y = -3/4 X + 3$ 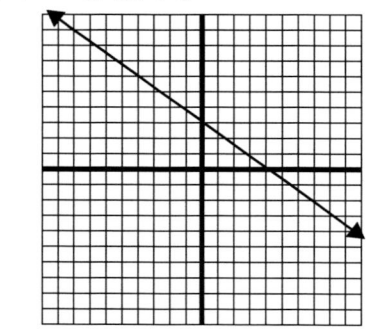
5. $Y = 1/3 X - 8/3$ 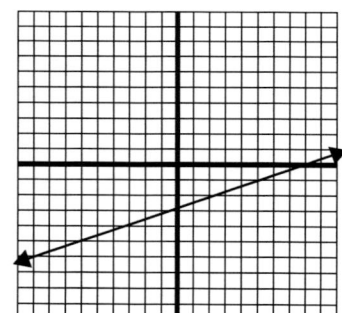
6. $Y = -3/2 X + 2$

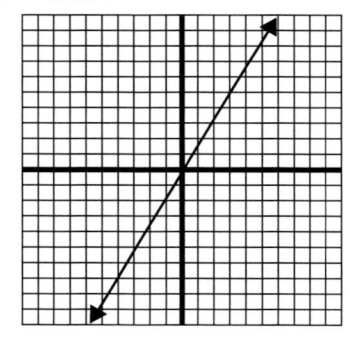

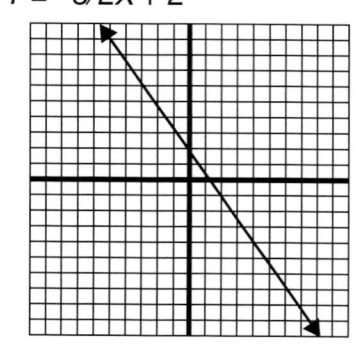

Answer Key   xxvii

7. $Y = 4/1 X + 2$

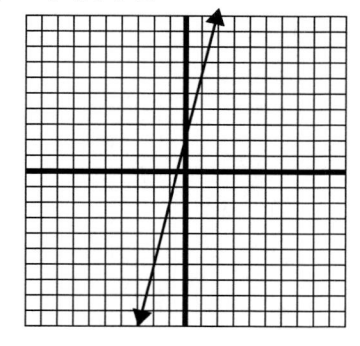

8. $Y = 1/1 X - 5$

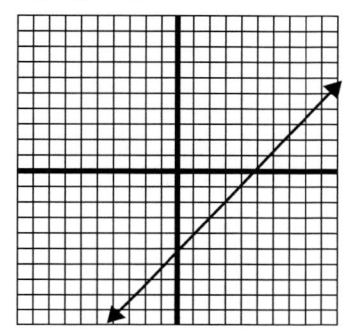

9. $Y = 3/2 X - 5/2$

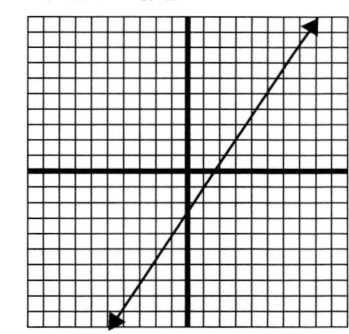

10. $Y = 3/2 X + 4$

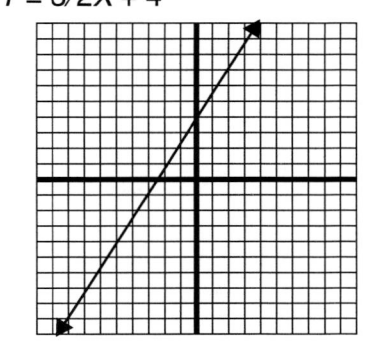

40. Finding Intersections by Plotting the Graphs

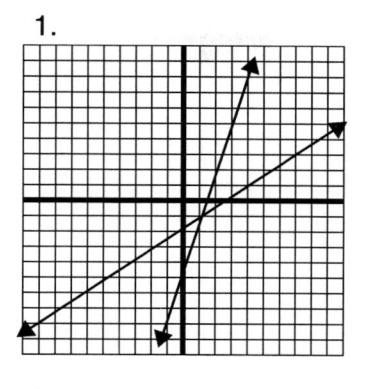

1.
$X \approx 1.5$
$Y \approx -.75$

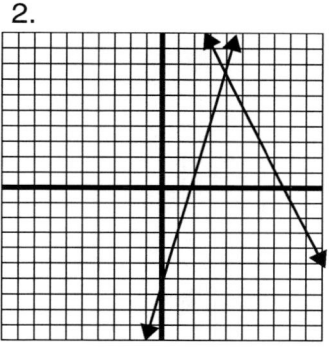

2.
$X \approx 3.5$
$Y \approx 7.5$

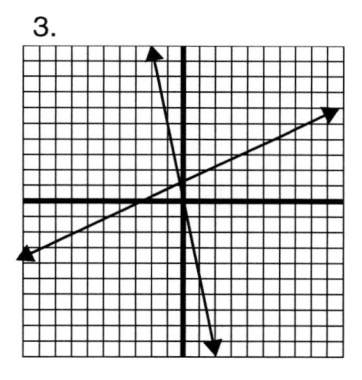

3.
$X \approx -.25$
$Y \approx 1$

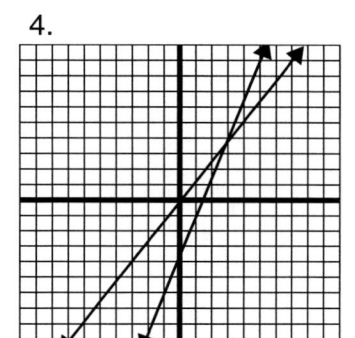

4.
$X \approx 2.75$
$Y \approx 3.25$

xxviii   Algebra Practice Exercises

5.

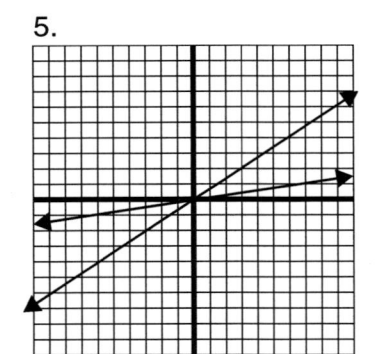

$X \approx 0$
$Y \approx 0$

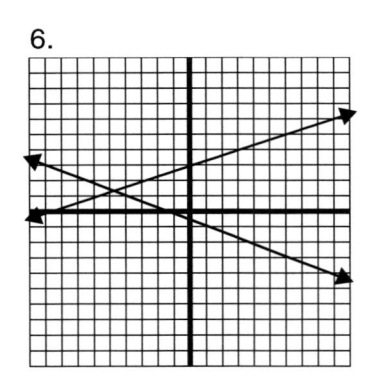

6.

$X \approx -5$
$Y \approx 1.5$

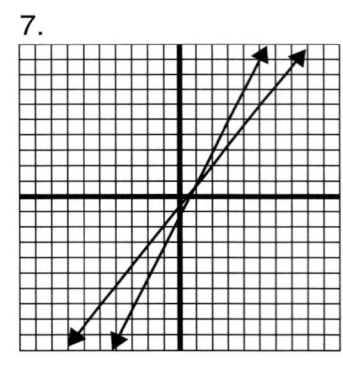

7.

$X \approx 0$
$Y \approx -1$

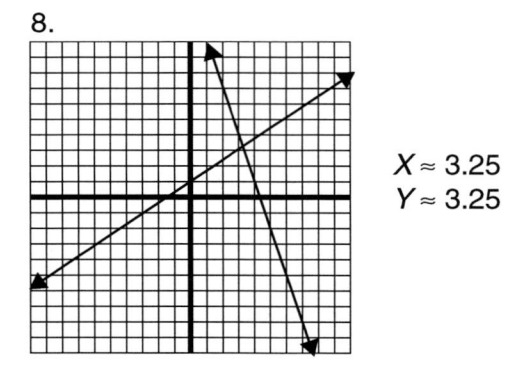

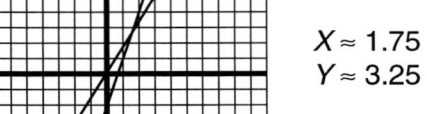

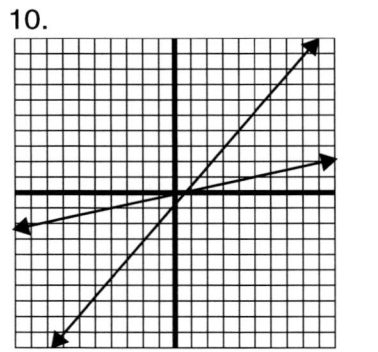

8.

$X \approx 3.25$
$Y \approx 3.25$

9.

$X \approx 1.75$
$Y \approx 3.25$

10.

$X \approx .5$
$Y \approx 0$

41. Solving Simultaneous Equations by the Substitution Method

1. (1, −1)
2. (−12, −41)
3. (3.5, 3)
4. (3, 4)
5. (1, $\frac{1}{7}$)
6. (6, 3)
7. (1, .5)
8. ($1\frac{1}{2}$, 2)
9. (1.2, .8)
10. (−2, −3)
11. Answers will vary. *Examples:* $2X + Y = 13$ and $3Y − X = 4$
12. Answers will vary. *Examples:* $3X − 3Y = 3$ and $Y − 3X = 3$

42. Solving Simultaneous Equations by the Elimination Method

1. (1, −1)
2. (−12, −41)
3. (3.5, 3)
4. (3, 4)
5. (1, $\frac{1}{7}$)
6. (6, 3)
7. (1, .5)
8. ($1\frac{1}{2}$, 2)
9. (1.2, .8)
10. (−2, −3)
11. (2.2, .4)
12. ($1\frac{3}{11}$, $1\frac{9}{11}$)

43. Solving Word Problems

1. Shawn and Ryan are 5 years old now.
2. Terri started with 54 marbles.
3. Ross is 34 years old now.
4. Susan can make 27 cabinets each week.
5. Adam has 48 cards.
6. Richard and Muriel have played 100 games so far.
7. Katie has 160 butterflies right now.
8. Margy is 4 years old now.
9. Steve bought 177 CD's.
10. Carol's car has 7,440 miles on it now.

xxx   Algebra Practice Exercises

44. Scientific Notation

1. 3.835×10^1
2. 4.78×10^1
3. 4.096×10^3
4. 1.6×10^1
5. 5.001×10^4
6. 8.2×10^{-2}
7. 1.048576×10^6
8. 7.600003×10^{-5}
9. .0034
10. 78,008,000
11. .00006701
12. 235,000,000
13. 7,501,000,000,000
14. 0.0000000000314
15. 3.5×10^1
16. 8.0×10^9
17. 7.0×10^7
18. 1.6×10^{-2}
19. 9.0×10^2
20. 2.8×10^0

45. Simplifying Rational Expressions

1. $\dfrac{1}{3AB}$
2. X^2
3. $\dfrac{-8J}{K}$
4. $\dfrac{U^3 V}{5}$
5. $(e+1) \times (f^2 - f + 1)$
6. r^2
7. $m^2 g^7$
8. $\dfrac{1}{x^2 Z^3}$
9. $\dfrac{m^{k+2}}{n^5}$
10. $\dfrac{k^2 m}{b^2}$
11. $\dfrac{2b^2}{A^5 D^3}$
12. $\dfrac{2H^3}{G^4}$
13. $\dfrac{w^2}{z^8}$
14. $\dfrac{m-3}{m+1}$
15. $\dfrac{c}{c+8}$
16. $\dfrac{t^2(t+2)}{t-1}$
17. $r + s$

46. Solving Simple Inequalities

1.
```
        <────────────────────○
<--|--|--|--|--|--|--|--|--|--|--|-->
  -2 -1  0  1  2  3  4  6  6  7  8  9
```

2.
```
                  ●───────────────→
<--|--|--|--|--|--|--|--|--|--|--|-->
   1  2  3  4  5  6  7  8  9 10 11 12
```

3.
```
              ○───────────────────→
<--|--|--|--|--|--|--|--|--|--|--|-->
   1  2  3  4  5  6  7  8  9 10 11 12
```

4.
```
         ●────────────────────────→
<--|--|--|--|--|--|--|--|--|--|--|-->
   0  1  2  3  4  5  6  7  8  9 10 11
```

5.
```
         ●────────────────────────→
<--|--|--|--|--|--|--|--|--|--|--|-->
   0  1  2  3  4  5  6  7  8  9 10 11
```

6.
```
              ●──────────────────→
<--|--|--|--|--|--|--|--|--|--|--|-->
  20 21 22 23 24 25 26 27 28 29 30 31
```

7.
```
   ←──────────────────○
<--|--|--|--|--|--|--|--|--|--|--|-->
  -5 -4 -3 -2 -1  0  1  2  3  4  5  6
```

8.
```
   ←──────────────────○
<--|--|--|--|--|--|--|--|--|--|--|-->
   1 1¹⁄₁₀ 1²⁄₁₀ 1³⁄₁₀ 1⁴⁄₁₀ 1⁵⁄₁₀ 1⁶⁄₁₀ 1⁷⁄₁₀ 1⁸⁄₁₀ 1⁹⁄₁₀ 2 2¹⁄₁₀
```

9.
```
   ←────────────────────────●|
<--|--|--|--|--|--|--|--|--|--|--|-->
  1⅓ 1⅔  2  2⅓ 2⅔  3  3⅓ 3⅔  4  4⅓ 4⅔  5
```

10.
```
   ←────────────────────────○
<--|--|--|--|--|--|--|--|--|--|--|-->
   0  1  2  3  4  5  6  7  8  9 10 11
```

Answer Key   xxxi

47. Solving Compound Inequalities

1. $Y<1$ or $Y\leq\frac{2}{3}$ (overlap)

2. $-5\frac{1}{2}<D<-2$

3. $K<1$ or $K>4$

4. $-5\leq V\leq-\frac{1}{2}$

5. $-9<Y<-2$

6. $-5\frac{1}{3}\leq T\leq-2$

7. $-5<B<2$

8. $-11\leq Y<10$

9. $-1\frac{1}{2}\leq P\leq 3\frac{1}{2}$

10. $\frac{5}{6}\leq H\leq 1$

48. Solving Compound Absolute Value Inequalities

1. $-\frac{1}{2} < J < 2$

2. $-1 < R < 7$

3. $X \leq -6$ or $X \geq 4$

4. $-1\frac{1}{2} < F < 3\frac{1}{2}$

5. $-1 \leq T \leq 5$

6. $-\frac{4}{5} < Y < 0$

7. $2\frac{1}{2} \leq P \leq 4\frac{1}{2}$

8. True for all T.

9. $-8 < Z < -2$

10. $-4\frac{1}{4} \leq A \leq 1\frac{1}{4}$

49. Graphing Linear Inequalities

1. 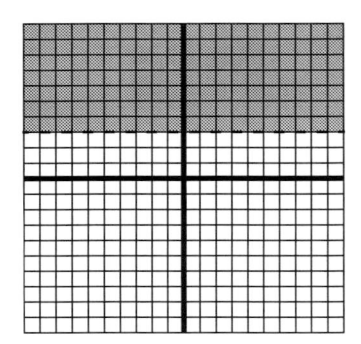
2. 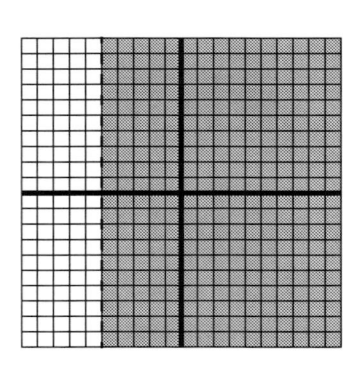
3. 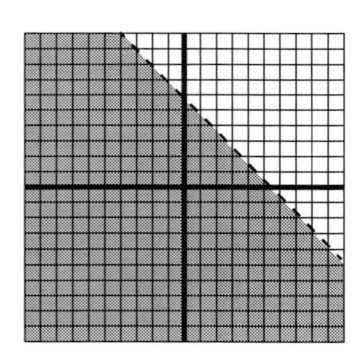
4. 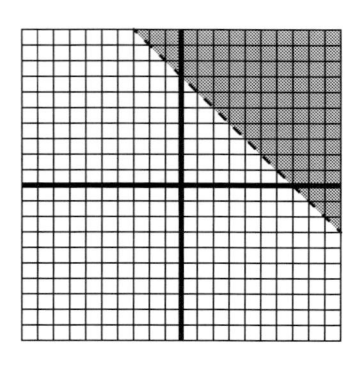
5. 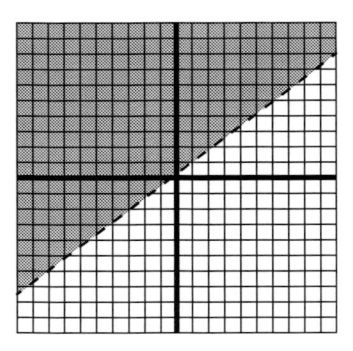
6. 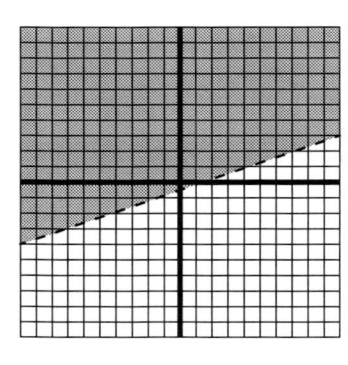
7. 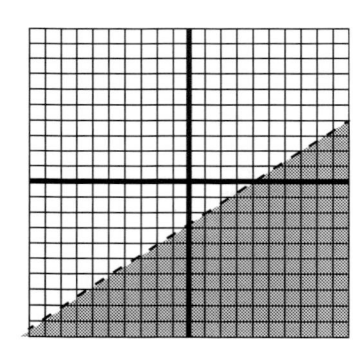
8.

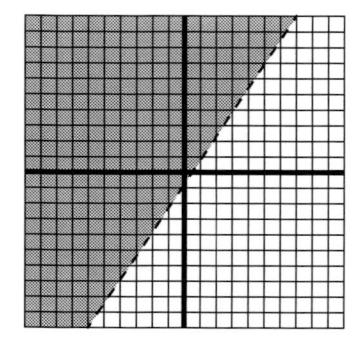

Answer Key xxxv

9.

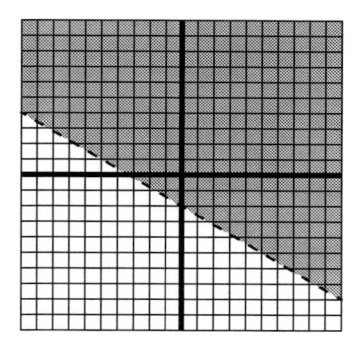

10.

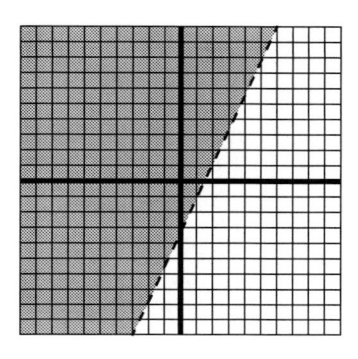

11. $Y < 6$

12. $X + Y \geq 0$

50. Graphing Systems of Linear Inequalities

1.

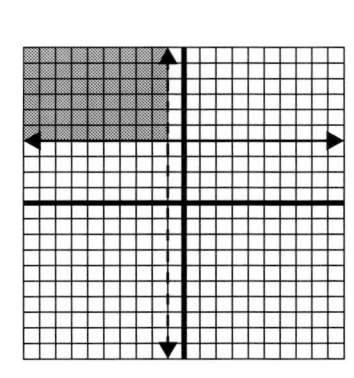

2.

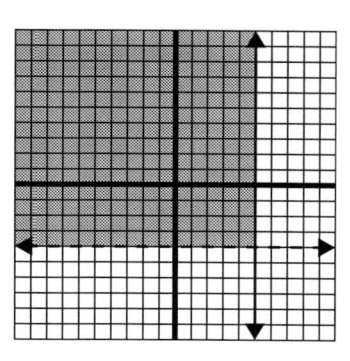

3.

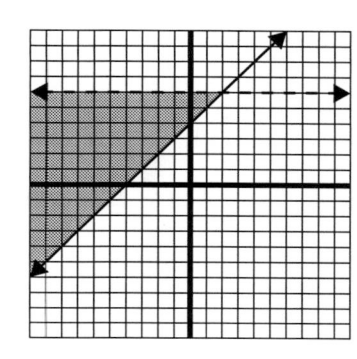

4.

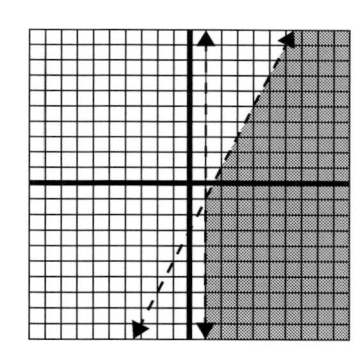

5.

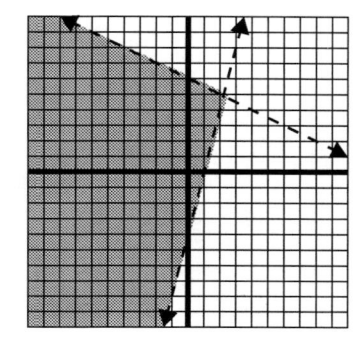

xxxvi   Algebra Practice Exercises

6.

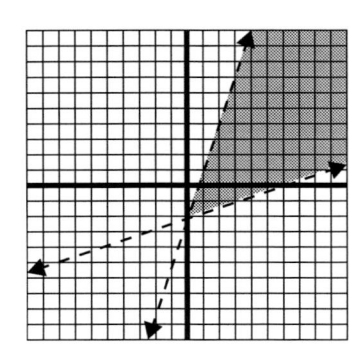

7.

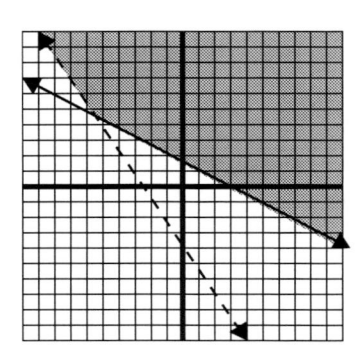

8.

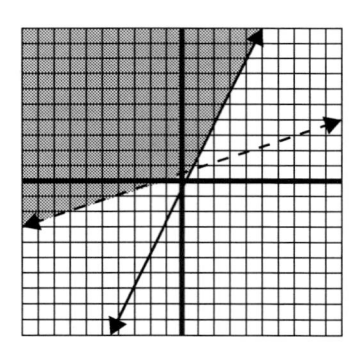

9.

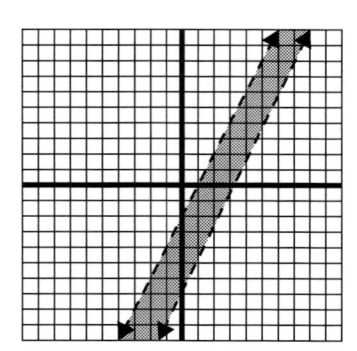

10.

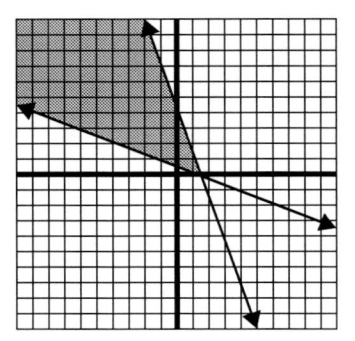

Name _____   Date _____

1. Identifying Algebra Vocabulary and Properties

1. Answer the following about the algebraic statement $4\,ax^2 + 3\,cxy$:

    (a) Name an element that appears in the statement. _____

    (b) What operation(s) are used? _____

    (c) What is a common factor of the terms? _____

    (d) What element might be called a coefficient? _____

2. Draw a line from each property in the left column to the correct algebraic statement in the right column.

    (a) Associative property   (i) $a = b$ and $b = c$ implies $a = c$

    (b) Distributive property   (ii) $(a + b) + c = a + (b + c)$

    (c) Identity element   (iii) $a + b = b + a$

    (d) Undefined operation   (iv) $a \times (b + c) = ab + ac$

    (e) Transitive property   (v) $e \times a = a$

    (f) Commutative property   (vi) $a \div 0$

3. What property or properties would allow you to calculate the answer to $97 \times 12 + 3 \times 12$ more easily? _____

4. What number is the identity for the operation "addition"? _____

5. Name an operation that is commutative. _____

6. Name an operation that is not commutative. _____

7. Name an operation that is associative. _____

8. Name an operation that is not associative. _____

9. According to the closure property, a set of numbers is "closed" under an operation, *, if for any a and b in the set, $a * b$ is also in the set. Is Z^+ (the set of positive integers) closed under division? If not, why not? _____

© 1996 J. Weston Walch, Publisher

Algebra Practice Exercises

Name _____ Date _____

2. Fractions, Decimals, and Percents

Convert the following as indicated:

1. $\frac{3}{5}$ to a decimal _____

2. .86 to a percent _____

3. .606 to a fraction _____

4. $\frac{1}{4}$ to a percent _____

5. 45% to a fraction _____

6. 57.5 % to a decimal _____

7. 86% to a fraction _____

8. $\frac{5}{6}$ to a decimal _____

9. .75 to a fraction _____

10. .07 % to a decimal _____

11. $\frac{1}{5}$ to a percent _____

12. .375 to a percent _____

13. .666–> to a fraction _____

14. $\frac{1}{7}$ to a decimal _____

15. $\frac{5}{4}$ to a percent _____

16. .272727–> to a fraction _____

17. $\frac{4}{9}$ to a percent _____

18. $\frac{3}{11}$ to a decimal _____

19. $\frac{5}{16}$ to a percent _____

20. .353535353535–> to a fraction _____

© 1996 J. Weston Walch, Publisher

Algebra Practice Exercises

Name _____ Date _____

3. Solving Percent Problems for the Part

Solve the following:

1. $W = .75 \times (96)$ _____
2. $X = .60 \times (120)$ _____
3. $Q = \frac{55}{100} \times (80)$ _____
4. N is $.25 \times (68)$ _____
5. G is $\frac{36}{100} \times (200)$ _____
6. K is $50\% \times (638)$ _____
7. X is $33\frac{1}{3}\% \times (99)$ _____
8. Y is 20% of 75? _____
9. M is 36% of 250? _____
10. What is 14% of 350? _____
11. What is 88% of 275? _____
12. What is 54% of 450? _____

13. Ryan has a marble collection. He recently calculated that he had 224 marbles and 75% of them were clear. How many of the marbles were clear? _____

14. On October 29, 1929, the Stock Market crashed. On the 28th, the Dow Jones Industrial Average was 260.64 points. On the 29th, it lost about 11.7% of its value. Approximately what amount of value was lost? _____

15. If 31.1% of all ninth graders require vision correction to see accurately, and there are currently about 243,000 ninth graders in America, how many ninth graders require vision correction? _____

© 1996 J. Weston Walch, Publisher   Algebra Practice Exercises

Name _____   Date _____

4. Solving Percent Problems for the Rate

Solve the following:

1. $64 = K \times 80$ _____

2. $27 = A \times (108)$ _____

3. $36 = T \times (300)$ _____

4. 14 is $\frac{I}{100} \times 56$ _____

5. 15 is $\frac{E}{100} \times 90$ _____

6. 16 is $B\% \times 256$ _____

7. $33\frac{1}{3}$ is $E\% \times 100$ _____

8. 18 is $T\%$ of 72? _____

9. 22 is $H\%$ of 110? _____

10. 14 is what % of 350? _____

11. 7 is what % of 8? _____

12. 200 is what % of 3,000? _____

13. In 1992, Bill Clinton received 44,909,000 votes for president, George Bush received 39,104,000 votes, and H. Ross Perot received 20,481,000 votes. Assuming no other candidate received a significant number of votes, what percentage of the vote did each candidate receive (to the nearest $\frac{1}{10}$ of 1%)? _____

14. The average Maine farmer is paid $0.60 per gallon of milk. The average American consumer pays $1.43 per half-gallon of milk. What percentage of the final supermarket price does the farmer receive on the sale of a half-gallon of milk? _____

15. A college recently graduated 330 students. Of these, 42 majored in mathematics. What percentage of the graduates were math majors? _____

© 1996 J. Weston Walch, Publisher

Algebra Practice Exercises

Name _____ Date _____

5. Solving Percent Problems for the Whole

Solve the following:

1. $18 = .75 \times X$ _____

2. $78 = .60 \times Z$ _____

3. $55 = 11/100 \times P$ _____

4. 77 is $.35 \times Y$ _____

5. 52 is $65/100 \times B$ _____

6. 990 is $72\% \times K$ _____

7. 56 is $66\frac{2}{3}\% \times M$ _____

8. 17 is 51% of J? _____

9. 47 is 70.5% of H? _____

10. 350 is 70% of what? _____

11. 12 is 8% of what? _____

12. 77 is 22% of what? _____

13. Terri and Larry have done some research about the twins they are having. They have learned that 2.3% of all pregnancies result in multiple births. If there is an average of 14,000 pregnancies in the state of Vermont each year, about how many result in multiple births? _____

14. Tyler collects postcards that are sent to him. He has noticed that he has 48 postcards depicting animals, which represent 60% of his collection. How many postcards are in his collection? _____

15. If there are 28,000 American 14-year-olds who have attended a rock concert, and this represents 12% of the population of 14-year-olds, about how many 14-year-olds are there in the United States? _____

© 1996 J. Weston Walch, Publisher   Algebra Practice Exercises

Name _____ Date _____

6. Solving Percent Problems

Solve the following:

1. $27 = .90 \times D$ _____

2. $78 = M \times 300$ _____

3. $Y = \frac{24}{100} \times 450$ _____

4. 66 is $.33\rightarrow \times Y$ _____

5. 75 is $\frac{K}{100} \times 160$ _____

6. J is $55\% \times 80$ _____

7. N is $33\frac{1}{3}\% \times 210$ _____

8. 34 is 51% of J? _____

9. 49 is $L\%$ of 50? _____

10. 35 is 140% of what? _____

11. What is 108% of 225? _____

12. 260 is what % of 52? _____

13. There are 36 different possible outcomes when two dice are rolled, and there are three different ways to roll a four. What percent of the time could you expect to roll a four, if you rolled two dice a large number of times? _____

14. Margy has learned that there are 65,000 registered owners of Jaguars™ in America. Her research shows that 96% of those owners are satisfied with their cars. How many owners are satisfied with their Jaguars? _____

15. The number of hot dogs sold at a minor league baseball game is about 22% of the number of fans in attendance. If 1,234 hot dogs were sold at a recent Portland Sea Dogs game, estimate the size of the crowd. _____

16. If the bill at a restaurant is $56.24 and you know the hospitality tax charged in this state is 8%, calculate the price of the meal without tax. _____

Name _____   Date _____

7. Order of Operations

Simplify the following:

1. 6 (3 + 5) = _____

2. 4 × 3 + 4 ÷ 2 = _____

3. 6 (2 + 3 ÷ 2) = _____

4. 17 − 1 (4 + 6) = _____

5. 4^3 + 2 − 10 ÷ 5 = _____

6. 4 + 2^3 − 11 = _____

7. 3 (2 − 1) (1 − 2) = _____

8. 2 (6 − 1) 1 + 6 = _____

9. $(3 + 4)^2$ − 12 × 2 ÷ 4 = _____

10. Use four 7's as your only digits, and any math symbols you want, to write a math statement that simplifies to 18. _____

Insert parentheses in the following equations to make them true:

(*Example:* Given: 2 1 + 1 3 − 1 7 = 35; insert parentheses to create: 2 1 + 1 (3 − 1) 7 = 35)

11. 3 2 − 1 3 + 2 1 = 21 _____

12. 3 2 − 1 3 + 2 1 = 30 _____

13. 3 2 − 1 3 + 2 1 = 114 _____

14. 3 2 − 1 3 + 2 1 = 40 _____

15. 3 2 − 1 3 + 2 1 = −31 _____

16. 3 2 − 1 3 + 2 1 = −5 _____

17. 3 2 − 1 3 + 2 1 = 15 _____

18. Create your own possible solution and challenge your neighbor:

    3 2 − 1 3 + 2 1 = _____

© 1996 J. Weston Walch, Publisher

Algebra Practice Exercises

Name _____   Date _____

8. Evaluation of Expressions

Find the value of the given expressions:

1. $3X - 2Y$, if $X = 4$ and $Y = 2$ _____

2. $12M - 13H$, if $M = 7$ and $H = 5$ _____

3. $4R + 5K$, if $R = 5$ and $K = -8$ _____

4. $3T - 2U$, if $T = -3$ and $U = -6$ _____

5. $7J - 8K$, if $J = -9$ and $K = -4$ _____

6. $-3[L + 7(3 - L)] + 4(L + 3)$, if $L = -2$ _____

7. $X^2 + 2YX + Y^2$, if $X = 3$ and $Y = -1$ _____

8. $EF^2 - 6E + 7F - 11$, if $E = 4$ and $F = -3$ _____

9. $(U - V)(U + V)$, if $U = 22$ and $V = 3$ _____

10. $3(K^2 - 7K) + 6T$, if $K = 2$ and $T = -7$ _____

11. $(4 - Y) - (5 - Y) - (6 - Y) + 4T$, if $Y = -3$ and $T = 0$ _____

12. $5[2(K - 2) - (3 - K)] + 6 - K$, if $K = -9$ _____

13. $M - 4[6(Z - 3) - (3 - M)]$, if $M = -3$ and $Z = 2$ _____

14. $F^2GH - FG^2H + FGH^3 - 4$, if $F = -1$, $G = -2$, and $H = 3$ _____

15. $6XY - X^2 + 5Y^2 - 7X$, if $Y = -2$ and $X = -3$ _____

16. $8[P(3 - K) - 2P + K] + 8P - 4K$, if $P = -2$ and $K = -1$ _____

17. $-LNC^2 + C(L + 3N) - N^2$, if $L = 3$, $N = 4$, and $C = -5$ _____

18. $-4PQ^2V + PV^2 - P^3 - V^3 - 5$, if $P = -2$, $Q = -3$, and $V = -1$ _____

© 1996 J. Weston Walch, Publisher   Algebra Practice Exercises

Name _____ Date _____

9. Solving Simple Equations

Solve for the unknown:

1. $X = 7 - 3$ _____

2. $12 = A + 7$ _____

3. $Y - 4 = 16$ _____

4. $2M - 3 = 11$ _____

5. $3E + 4 = 25$ _____

6. $13 = 7G - 8$ _____

7. $5A = 7 + 28$ _____

8. $18 - 5N = 3$ _____

9. $6B - 13 = 53$ _____

10. $7U - 3 = U + 57$ _____

11. $3T + 9 = 7T - 35$ _____

12. $6(2L - 3) = 7L + 17$ _____

13. $3E - 16 = 5(2E + 15)$ _____

14. $2R + 7 - R = 7R - 53$ _____

15. $6I - 78 = 15 - 2I$ _____

16. $53 + 4L = 14 - 3L$ _____

17. $43 - 17Y + 6 = 7Y - 18$ _____

© 1996 J. Weston Walch, Publisher   *Algebra Practice Exercises*

10. Solving Absolute Value Equations

Solve for the unknown. Give numberline and algebraic solutions:

1. $|X - 4| = 12$
2. $|B + 5| = 6$
3. $|E - 6| = 22$
4. $|16 - T| = 5$
5. $14 = 2|H - 3|$
6. $|2A - 3| = 27$
7. $|8 - 2N| = 3N$
8. $|3Y - 4| = 5Y$
9. $|3 - 7| = |I + 3|$
10. $|4V - 5| = 15$
11. $|4 - E| = -18$
12. $4|S - 3| = 19$
13. $4|I - 2| = |15 - 22|$
14. $|L + 4| = |5 - 2L|$
15. $|Y - 3| = |2Y - 8|$
16. $||T + 1| - 4| = 12$
17. $||C - 3| - 2| = 7$

Name _____ Date _____

11. Exponents

Simplify the following:

1. $(-3A^4B^3)(-4A^2B^5)$ _____

2. $(2X^2Y^3Z)(-5X^4YZ^6)$ _____

3. $(5X^3Y^6)^3$ _____

4. $(4F^4G^5H^3)^4$ _____

5. $\dfrac{34U^3D^{-5}}{51U^5D^2}$ _____

6. $\dfrac{25F^2G^{-5}R^4}{15F^{-3}G^{-2}R^5}$ _____

7. $\dfrac{32M^3N^0P^{-4}}{8M^{-2}P}$ _____

8. $\dfrac{(44Z^4X^3Y^7)^0}{55Z^3X^{-2}}$ _____

9. Explain why $(RS)^x$ is equivalent to R^xS^x _____

10. Explain why $(W^x)^y$ is equivalent to W^{xy} _____

Simplify:

11. $\dfrac{(CD)^{-3}}{C^5} + \dfrac{1}{C^4D^{-4}}$ _____

12. $\dfrac{4G^4}{H^{-4}} - \dfrac{(H^{-3})^{-2}}{G^{-5}}$ _____

13. $8^{2/3}$ _____

14. $81^{3/4}$ _____

15. $5^{1-\sqrt{2}} \cdot 5^{1+\sqrt{2}}$ _____

16. $7^{e-4} \cdot 7^{e+4}$ _____

© 1996 J. Weston Walch, Publisher   *Algebra Practice Exercises*

Name _____ Date _____

12. Operations with Negative Numbers

Simplify the following:

1. 12 + –7 _____
2. 13 + –3 _____
3. –6 + –5 _____
4. 10 – 22 _____
5. 11 – –8 _____
6. –3 – –9 _____
7. –2 (–9) _____
8. –4 (–6 + –3) _____
9. 7 – –5 + 4 (9 – –7) _____
10. 7 – 8 (3 – –4) _____
11. –19 + 23 – –9 – 3 (2 + 3) _____
12. –14 – (–7 + –3 – –9) + –11 _____
13. –17 – –27 + –3 – +7 _____
14. –26 + 28 – (16 – 23) + (–3 – 5) _____
15. –45 – –117 + 41 + –89 – 32 (–9 – –9) _____
16. –90 + 45 – 34 + 17 (23 – 19) _____
17. –11 [–3 (–7 – –15) + 18] _____

18. Record the following situation numerically. Peter is getting into a hot-air balloon. He has 32 sandbags with him and has calculated that each of them will reduce his altitude by 4 feet. He decides to leave 18 bags on the ground.

_____

19. Write a story problem that would use the calculation –3 (54 – 48) _____

© 1996 J. Weston Walch, Publisher

Algebra Practice Exercises

Name _____ Date _____

13. Distribution and Double Distribution

Multiply the following problems out:

1. $4(3X + 5Y)$ _____
2. $3B(4X - 7)$ _____
3. $6C(2 + 3D)$ _____
4. $3CR(2C - 3R)$ _____
5. $5X(3X^2 - 2X - 1)$ _____
6. $7Y^2Z(3Z - 5Y^2)$ _____
7. $4D(3D - 7E + 2D^2)$ _____
8. $5N(7 - 3N - 2N^2)$ _____
9. $(X - 3)(X + 2)$ _____
10. $(Z - 7)(Z - 4)$ _____
11. $(N + 1)(N + 12)$ _____
12. $(K - 4)(K + 6)$ _____
13. $(2X - 5)(X + 3)$ _____
14. $(4N - 3)(M + 1)$ _____
15. $(10Q - 2)(Q - 3)$ _____
16. $(2W - 3V - 1)(W + 7)$ _____
17. $(2C - 3D + 2)(F + 1)$ _____
18. Explain why the answers to problems 16 and 17 have different numbers of terms, even though the problems have the same number of terms. _____

_____

_____

© 1996 J. Weston Walch, Publisher   Algebra Practice Exercises

Name _____ Date _____

14. Combining Like Terms

Simplify and combine like terms in the following problems. (Be careful. Not all can be simplified.)

1. $X^2 - 2X - 3X + 5$ _____

2. $K^2 - 7K + 3K - 21$ _____

3. $15X - 3Z + 2X$ _____

4. $13Z + 4Z^2 - 7Z$ _____

5. $3XY - 2ZY + 2XZ - XY$ _____

6. $14AB - 12B^2 - 7A + 3B^2 - 7AB$ _____

7. $X^2 - 2Y^2 + 3X - 2Y - 3X^2 + 7 - 2XY$ _____

8. $\dfrac{3X^2}{X} - 2X + 7 - 3XY - \dfrac{2X}{X}$ _____

9. $DV - 6V^2D + 7VD - 3$ _____

10. $43MN + 77N^2M - 23M^2N - 2M^2$ _____

11. $X^3YZ - 7X^2Y^2Z + 3XYZ - 6XYZ^3$ _____

12. $7T^2H + 18TH - 6TH^2 + 12HT$ _____

13. $3X(2X - 4Y) + 5Y(3X - Y)$ _____

14. $(2A - 3B)(3A + 2B)$ _____

15. $14VW(6V - 3W + 1) + 4W(3WV - 2V)$ _____

16. $\dfrac{3G^2H^2 - 46GH^3}{GH} + \dfrac{3G^2H^2 - 46GH^3}{H^2}$ _____

17. $(2C - 3D + 2)(C - 5D + 1)$ _____

18. $(3WY - 7Y + 4)(2WY - 3W - 2)$ _____

© 1996 J. Weston Walch, Publisher   Algebra Practice Exercises

Name _____ Date _____

15. Divisibility Rules

The common rules for divisibility are as follows:

(2) A number is divisible by 2 only if its last digit is even.
(3) A number is divisible by 3 only if its digits add up to a number divisible by 3.
(4) A number is divisible by 4 only if its last two digits are divisible by 4.
(5) A number is divisible by 5 only if its last digit is a 0 or a 5.
(6) A number is divisible by 6 only if its digits add up to a number divisible by 3 and its last digit is even.
(7) A number is divisible by 7 only if the alternating sum of the three digit numbers made from its digits as separated by the place value commas is divisible by 7. Example of an alternating sum: 307,739,901 would be
$$307 - 739 + 901 =$$
(8) A number is divisible by 8 only if its last three digits are divisible by 8.
(9) A number is divisible by 9 only if its digits add up to a number divisible by 9.
(10) A number is divisible by 10 only if its last digit is a 0.
(11) A number is divisible by 11 only if the alternating sum of the three digit numbers made from its digits as separated by the place value commas is divisible by 11. See **(7)** for an example of an alternating sum.

Determine the divisibility of the following:

1. Is 349 divisible by 3?

2. Is 24,698 divisible by 2?

3. Is 376,343 divisible by 11?

4. Is 578 divisible by 6?

5. Is 6,525 divisible by 5?

6. Is 7,865,900 divisible by 10?

7. Is 300,492 divisible by 7?

8. Is 18,261 divisible by 9?

9. Is 312,987,654 divisible by 4?

10. Is 54,354 divisible by 8?

11. Is 765,934,435 divisible by 7?

12. Is 295,987,102 divisible by 3?

13. Is 5,498,230 divisible by 5?

14. Is 49,378,278 divisible by 6?

15. Is 763,934 divisible by 4?

16. Is 827,973 divisible by 9?

17. A number is divisible by 21 only if it is divisible by 7 *and* by 3. Is 3,197,453 divisible by 21?

© 1996 J. Weston Walch, Publisher   *Algebra Practice Exercises*

Name _____   Date _____

16. Factoring to Primes

Factor the following numbers completely and write them in exponential form. (Be careful. At least one of them cannot be factored.)

1. 216 _____
2. 2,520 _____
3. 5,625 _____
4. 306 _____
5. 256 _____
6. 1,134 _____
7. 2,550 _____
8. 1,078 _____
9. 1,859 _____
10. 11,088 _____
11. 443 _____
12. 1,513 _____
13. 791 _____
14. 789 _____
15. 3,840 _____
16. 11,907 _____
17. 10,647 _____
18. 33,075 _____
19. 93,925 _____
20. 2,097,152 _____

Name _____   Date _____

17. Finding the Least Common Multiples

Find the smallest number that is a multiple of each of the following groups of numbers:

1. 34 and 51 _____
2. 18 and 24 _____
3. 21 and 28 _____
4. 72 and 18 _____
5. 144 and 42 _____
6. 6 and 7 and 14 _____
7. 256 and 144 and 64 _____
8. 225 and 55 and 27 _____
9. 128 and 7 and 28 _____
10. 169 and 52 and 4 _____
11. 196 and 42 and 3 _____
12. 63 and 441 and 14 _____
13. 256 and 13 and 52 _____
14. 4,096 and 6 and 320 _____
15. 108 and 18 and 12 _____
16. 15 and 13 and 14 _____
17. 31 and 7 and 12 _____
18. 1,296 and 15 and 80 _____
19. If 495 is the LCM of 11, 5, and another number, what might that number be? ____
20. If 1,386 is the LCM of 6, 14, and another number, what might that number be? __

© 1996 J. Weston Walch, Publisher   17   Algebra Practice Exercises

Name _____ Date _____

18. Finding Common Factors

Find the common factors of the following:

1. 34 XY and 24 XZ _____

2. 35 ABC and 28 BCD _____

3. 21 D^2E and 13 DE^2 _____

4. $54XY^2Z^3$ and 9 X^2Y^2Z _____

5. $144EF^4$ and $42E^3F^2$ _____

6. 6 R^3S^5 and $18R^2X^3$ and $21S^3X^4$ _____

7. $25XM$ and $125X^3M^6$ and $75X^2M^9$ _____

8. 14 CV^2 and $35C^3S^5$ and $10VS^3$ _____

9. $45MP^3$ and $81M^6P^3$ and $27PM^3$ _____

10. $169UV^2M$ and $65V^2UM$ and $91MU^2V$ _____

Write the following problems in fully factored form:

11. 28 $T^2EC^3 - 35\ CT^2H^4$ _____

12. $36YX^2 - 14XY^3 - 24XY$ _____

13. 75 $X^4D^2F + 225X^2D^5F^4 - 80XD^3F^2$ _____

14. $441J^4K^5L - 63J^2LK^2J + 105K^7L^3J^3$ _____

15. $256G^7H^5 + 64\ P^2G^6H^5 - 1024G^4H^7D^{10}$ _____

16. $504R^4G^2P - 18\ P^2GH^5 + 77G^2HR^6$ _____

17. 360 $S^5CT^2 + 135\ T^3S^2C - 60\ C^6T^5S$ _____

© 1996 J. Weston Walch, Publisher

Algebra Practice Exercises

Name _____   Date _____

19. Factoring Out the Greatest Common Factor

Factor the following completely. (One of them cannot be factored. Why?)

1. $2X^2 - 3X^3$ _____

2. $15a^2 + 5b^2$ _____

3. $12W^2 + 3W^3$ _____

4. $7m^2n^3 - 21m^3n^2$ _____

5. $t^6s^5 + s^4t^4$ _____

6. $121P^3Q^2 + 169D^3Q$ _____

7. $X^3Y^3Z - XY^2Z^2$ _____

8. $11a^4x^2 - 88a^3x^3$ _____

9. $144x^2y^5 - 36x^3y$ _____

10. $26d^2h^2 - 39d^2h^3$ _____

11. $7m^2n^3 - 21m^3n^2 + 63m^2n^2$ _____

12. $121P^3Q^2 + 165D^3Q + 77X^3Q^3$ _____

13. $17X^5Y^4 - 15X^5Z^3 + 13Y^2Z^3$ _____

14. $3X^3Y^3Z^3 - 2X^2Y^2Z^2 - XYZ$ _____

15. $a^3b^2c - a^2bc^3 + ab^3c^2$ _____

16. $14X^5 - 9Y^3 + 4X^5$ _____

17. $16\beta^3©^2 + 18\beta^3©^4 + 8©^4\beta^2$ _____

18. $1650X^2Y^3 + 250X^3Y - 850X^2Y$ _____

19. $1048576X^3 + 1024XY^2 - 32768X^2Z$ _____

20. Never Neat Nellie has factored a problem correctly (but illegibly) as follows:
    $X^2Y^3(X^\otimes - Y^\otimes + 1)$

    If both of Nellie's illegible exponents are positive and the original problem contained

    no exponent greater than 4, what might the problem have been? _____

_____

© 1996 J. Weston Walch, Publisher   Algebra Practice Exercises

Name _____ Date _____

20. Multiplying Binomials

Multiply out the following binomials:

1. $(X + 1)(X - 7)$ _____
2. $(W + 5)(W + 6)$ _____
3. $(K - 5)(K - 8)$ _____
4. $(A + 6)(A - 2)$ _____
5. $(T + 3)(T - 3)$ _____
6. $(H - 5)(H - 5)$ _____
7. $(E + 12)(E - 9)$ _____
8. $(R + 11)(R - 7)$ _____
9. $(J + 8)(J - 8)$ _____
10. $(N + 5)(N - 6)$ _____
11. $(E - 4)(E - 4)$ _____
12. $(C + 1)(2C - 3)$ _____
13. $(3H - 2)(H - 2)$ _____
14. $(2A + 3)(3A - 2)$ _____
15. $(3S - 4)(4S - 5)$ _____
16. $(3E + 1)(E - 9)$ _____
17. $(6J + 5)(5 - 2J)$ _____
18. $(3L - 7)(3L + 7)$ _____
19. $(5Y + 3)(5Y + 3)$ _____
20. $(9T - 4)(9T + 5)$ _____

© 1996 J. Weston Walch, Publisher

Algebra Practice Exercises

21. Factoring by Association

Factor completely:

1. $XY + 2Y + 3X + 6 = ($ _____ $)($ _____ $)$

2. $AB + 4B - 6A - 24 =$ _____

3. $2MX - 10M + 3X - 15 =$ _____

4. $24q - t + 4qt - 6 =$ _____

5. $pr + 4p - 3r - 12 =$ _____

6. $3LX - 4X + 21L - 28 =$ _____

7. $2WV + 6W - 6V - 18 =$ _____

8. $6K^3 + 3K^2 - 14K - 7 =$ _____

9. $12Z^3 + 24Z^2 - 14Z - 28 =$ _____

10. $ac + bc + c^2 - 6a - 6b - 6c =$ _____

11. $2mn + 2n^2 - 6n - m^2 - mn + 3m =$ _____

12. $4XYZ - 2Z^2 - WZ + 2WXY =$ _____

Never Neat Nellie is still having penmanship difficulties. She has completed the following correctly, so find the values of all the illegible marks:

13. $12🍎K - 28X + 9K - 🍎🍎 = (3K - 7)(4X + 🍎)$ _____

14. $2🍎W - WY + 3Y - 🍎X = (2X - Y)(🍎 - 🍎)$ _____

15. $🍎Z^2 - 6CZ - 2Z + 🍎🍎🍎 - 12🍎 - 🍎 = (4Z - 3🍎 - 1)(2🍎 + 4)$ _____

Name _____   Date _____

22. Factoring Trinomials by Undoing Distribution

Factor the following trinomials completely into binomial factors. (One of them cannot be factored. Why?)

1. $X^2 + 9X + 20 = ($ _____ $)($ _____ $)$

2. $Z^2 + 12Z + 32 =$ _____

3. $A^2 + 18A + 81 =$ _____

4. $m^2 - 18m + 81 =$ _____

5. $Q^2 + 18Q + 56 =$ _____

6. $L^2 - 15L + 56 =$ _____

7. $N^2 - 15N - 16 =$ _____

8. $C^2 + 6C - 16 =$ _____

9. $T^2 - 6T - 27 =$ _____

10. $Y^2 + 12Y + 27 =$ _____

11. $L^2 - 12L - 64 =$ _____

12. $E^2 + 30E - 64 =$ _____

13. $R^2 + 30R + 81 =$ _____

14. $M^2 + 1M + 1 =$ _____

15. $A^2 - 2A + 1 =$ _____

16. $N^2 + 3N + 2 =$ _____

17. $♠^2 - 13♠ + 30 =$ _____

Never Neat Nellie has copied three problems from the board and written her answers. The teacher believes that Nellie has answered each problem correctly. What were Nellie's problems and correct answers?

18. $G^2 - 10⊗ + ⊗ = (G - 7)(⊗ - ⊗)$ _____

19. $⊗^2 + 12H + ⊗ = (H + 3)(⊗ + ⊗)$ _____

20. $⊗^2 - ⊗K + ⊗ = (K ⊗ 7)(⊗ ⊗ 3)$ _____

© 1996 J. Weston Walch, Publisher   Algebra Practice Exercises

23. Solving Equations by Factoring and the Difference of Two Squares

Factor the following:

1. $X^2 + 18X - 40 = ($ _____ $)($ _____ $)$
2. $a^2 - 12a + 35 =$ _____
3. $y^2 - 2xy + x^2 =$ _____
4. $y^2 - 0y - 16 =$ _____
5. $z^2 - 121 =$ _____
6. $H^2 - 49 =$ _____
7. $F^2 - 576 =$ _____
8. Why can't $T^2 + 225$ be factored the same way as problems 5, 6, and 7 above?

   _____

9. $z^4 - 16 =$ _____
10. $G^4 - 81 =$ _____

Solve by factoring:

11. $m^2 - 3m - 10 = 0$ _____
12. $z^2 - 11z + 30 = 0$ _____
13. $e^2 - 10e + 25 = 0$ _____
14. $e^2 - 0e - 25 = 0$ _____
15. $e^2 - 25 = 0 =$ _____
16. Compare and contrast problems 14 and 15 above. _____

    _____

17. $M^4 - 64 = 0$ _____

24. Solving Quadratic Equations by Factoring

Solve for X by factoring:

1. $X^2 - 5X - 14 = 0$ _____
2. $X^2 + 7X + 10 = 0$ _____
3. $X^2 - 3X - 18 = 0$ _____
4. $X^2 + 4X - 12 = 0$ _____
5. $X^2 - 12X + 27 = 0$ _____
6. $X^2 - 7X + 12 = 0$ _____
7. $X^2 + 8X + 16 = 0$ _____
8. $X^2 + 0X - 36 = 0$ _____
9. $X^2 + 14X + 24 = 0$ _____
10. $X^2 - 18X + 81 = 0$ _____
11. $X^2 - 18X + 56 = 0$ _____
12. $X^2 - 18X - 40 = 0$ _____
13. $X^2 - 18X + 72 = 0$ _____
14. $X^2 + 18X + 17 = 0$ _____
15. $X^2 + 18X + 80 = 0$ _____
16. $X^2 - 18X + 77 = 0$ _____

Name _____ Date _____

25. A Variety of Factoring Strategies

Factor completely (difference of two squares):

1. $X^2 - 25$ _____
2. $4W^2 - 81$ _____
3. $M^4N^4 - U^2$ _____
4. $T^2 - 8$ _____
5. $P^4 - 36$ _____
6. $G^2 + 16$ _____
7. $Z^{16} - 1$ _____
8. Two of the above problems cannot be factored by the difference of two squares. Why not?
   _____

Factor completely (difference of two cubes):

9. $H^3 - 8$ _____
10. $125 - Y^3$ _____
11. $A^6 - 27$ _____
12. $216 - 343R^3$ _____

Factor completely (sum of two cubes):

13. $X^3 + 1$ _____
14. $K^3 + 64$ _____
15. $M^6 + 125$ _____
16. $1331T^3 + 343$ _____

Factor completely (by grouping):

17. $2B - AB + 6 - 3A$ _____
18. $X^2 + YX - 3X - 3Y$ _____
19. $2ef + 12e - 3f - 18$ _____
20. $X^2 - 4X + 4 - 4Y^2$ _____

26. Solving Complex Equations

Solve for the unknown:

1. $X - [\,3(2 - X) - 1\,] = 7$ _____

2. $(4 - Y) - (5 - Y) - (6 - Y) = 5Y - 18$ _____

3. $-3[\,G + 7(3 - G)\,] + 4(G + 3) = 7$ _____

4. $.5[\,2(K - 2) - (3 - K)\,] = 6 - K$ _____

5. $M - 4[\,6(M - 3) - (3 - M)\,] = 1$ _____

6. $3\,|V| - V + 3 = 2\,|V| - 1$ _____

7. $7U - 13 = 5(U - 6) - 17U - 12$ _____

8. $8[\,(t - 7) - (5 - 2t)\,] - 13 = 5t$ _____

9. $(5 - |G|) + 7 = 4 - 3G$ _____

10. $-\{\,7 - [\,8 - (9 - 2C)\,]\,\} = 7(2C - 3)$ _____

11. $13[\,T - 3(2 - 2T) - 11\,] = 14(3 - T) + 17T$ _____

12. $5(r - 3) - 4(3 - r) + 7(2r - 3) = 11(r - 3)$ _____

13. $3J - 2(9 - 2.5J) + 6.5 = 4.5 - 3J$ _____

14. $7.6A - 3.4 + 2(2.3 - 3.2A) = 9.8 - 1.2A$ _____

15. $6.2(.9E - .7) - (.3E - 1.2) = 4.5(3E - 2.1)$ _____

Algebra Practice Exercises

Name _____   Date _____

27. Simplifying Radicals

Simplify the following:

1. $\sqrt{576}$ _____
2. $\sqrt{729}$ _____
3. $\sqrt{162}$ _____
4. $\sqrt{605}$ _____
5. $\sqrt{1050}$ _____
6. $\sqrt{512}$ _____
7. $5\sqrt{48}$ _____
8. $\dfrac{2\sqrt{3}}{\sqrt{7}}$ _____
9. $\sqrt{\left(\dfrac{4}{9}\right)}$ _____
10. $\dfrac{5\sqrt{50}}{\sqrt{8}}$ _____
11. $4\sqrt{(2.5)}$ _____
12. $\dfrac{8\sqrt{4}}{4\sqrt{8}}$ _____
13. $\dfrac{15\sqrt{20}}{3\sqrt{5}}$ _____
14. $3\sqrt{5} - 2\sqrt{8} + 4\sqrt{5}$ _____
15. $12\sqrt{\left(\dfrac{243}{16}\right)}$ _____
16. $\sqrt{\left(\dfrac{7}{8}\right)} - \sqrt{\left(\dfrac{8}{7}\right)}$ _____
17. $(2\sqrt{5} - 5\sqrt{3})(3\sqrt{15} - 2\sqrt{6})$ _____
18. $(3\sqrt{3} - 5\sqrt{6})(4\sqrt{12} + 2\sqrt{18})$ _____
19. $\dfrac{5}{1+\sqrt{5}}$ _____
20. $\dfrac{7}{\sqrt{12}+3}$ _____

© 1996 J. Weston Walch, Publisher   *Algebra Practice Exercises*

28. The Pythagorean Theorem

Find the unknown lengths of the sides in the following triangles:

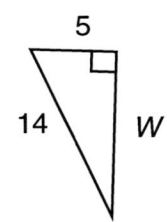

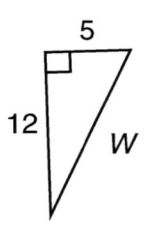

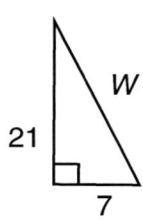

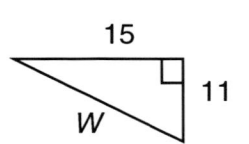

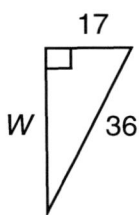

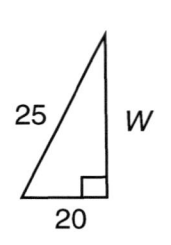

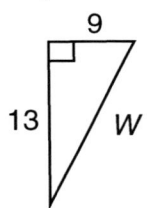

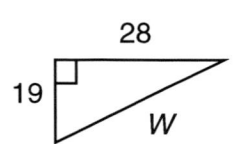

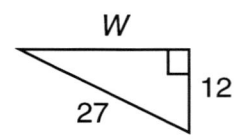

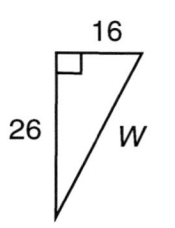

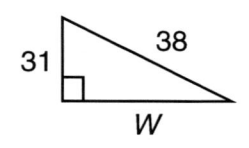

1. Legs 6 and 8, hypotenuse W.

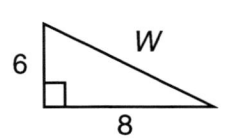

2. Legs 5 and 12, hypotenuse W.

3. Legs 15 and 11, hypotenuse W.

4. Leg 20, hypotenuse 25, other leg W.

5. Legs 19 and 28, hypotenuse W.

6. Legs 16 and 26... leg W with hypotenuse 26 and other leg 16. (W is leg; 26 hypotenuse, 16 leg)

7. Leg 5, hypotenuse 14, other leg W.

8. Legs 21 and 7, hypotenuse W.

9. Leg 17, hypotenuse 36, other leg W.

10. Leg 9, hypotenuse 13, other leg W.

11. Leg 12, hypotenuse 27, other leg W.

12. Leg 31, hypotenuse 38, other leg W.

Name _____ Date _____

29. Solving Quadratic Equations by Completing the Square

Solve for X by completing the square:

1. $X^2 - 4X - 12 = 0$ _____
2. $X^2 + 8X + 12 = 0$ _____
3. $X^2 - 6X + 8 = 0$ _____
4. $X^2 + 6X + 8.75 = 0$ _____
5. $X^2 - 12X + 35 = 0$ _____
6. $X^2 - 3X - 18 = 0$ _____
7. $X^2 + 5X + 6 = 0$ _____
8. $X^2 + 9X - 36 = 0$ _____
9. $X^2 + 4X - 9 = 0$ _____
10. $X^2 - 6X + 6 = 0$ _____
11. $X^2 - 8X - 5 = 0$ _____
12. $X^2 - 3X + 1.5 = 0$ _____
13. $X^2 + 7X + 2 = 0$ _____
14. $2X^2 + 6X + 2 = 0$ _____
15. $3X^2 - 9X - 6 = 0$ _____
16. $2X^2 + 5X - 7 = 0$ _____

Name _____ Date _____

30. Solving Quadratic Equations by the Quadratic Formula

Solve for X by the quadratic formula. (Be careful. One has no solution.)

1. $X^2 - X - 12 = 0$ _____

2. $X^2 + 8X + 15 = 0$ _____

3. $X^2 - 4X + 3 = 0$ _____

4. $X^2 + 6X + 6.75 = 0$ _____

5. $X^2 - 3X + 2.25 = 0$ _____

6. $X^2 - 5X - 7 = 0$ _____

7. $X^2 + 13X + 6 = 0$ _____

8. $X^2 + 7X + 9 = 0$ _____

9. $3X^2 + 4X - 7 = 0$ _____

10. $5X^2 - 12X + 5 = 0$ _____

11. $2X^2 - 7X - 14 = 0$ _____

12. $4X^2 - 12X + 6 = 0$ _____

13. $7X^2 + 14X + 14 = 0$ _____

14. $2X^2 + 12X + 7 = 0$ _____

15. $3X^2 - 10X - 2 = 0$ _____

16. $6X^2 + 14X - 9 = 0$ _____

© 1996 J. Weston Walch, Publisher   Algebra Practice Exercises

Name _____ Date _____

31. Solving Quadratic Equations

Solve for X by the method of your choice. (Be careful. At least one has no solution.)

1. $X^2 - X - 12 = 0$ _____

2. $X^2 + 8X + 15 = 0$ _____

3. $X^2 - 4X + 3 = 0$ _____

4. $X^2 + 6X + 6.75 = 0$ _____

5. $X^2 - 3X + 2.25 = 0$ _____

6. $X^2 - 5X - 7 = 0$ _____

7. $X^2 + 13X + 6 = 0$ _____

8. $X^2 + 7X + 9 = 0$ _____

9. $3X^2 + 4X - 7 = 0$ _____

10. $5X^2 - 12X + 5 = 0$ _____

11. $2X^2 - 7X - 14 = 0$ _____

12. $4X^2 - 12X + 6 = 0$ _____

13. $7X^2 + 14X + 14 = 0$ _____

14. $2X^2 + 12X + 7 = 0$ _____

15. $3X^2 - 10X - 2 = 0$ _____

16. $6X^2 + 14X - 9 = 0$ _____

32. The Cartesian Coordinate System

Place the following points on the Cartesian coordinate graph below:

1. (3, 4)
2. (5, 1)
3. (4, –3)
4. (–2, 3)
5. (2, 0)
6. (7, –2)
7. (–6, 8)
8. (–1, 0)
9. (–4, –5)
10. (–8, –9)
11. (4.5, 6)
12. (–2, 5.5)
13. (0, 0)
14. (–8.5, 9)
15. (–1.5, –9.5)
16. (6, –.5)

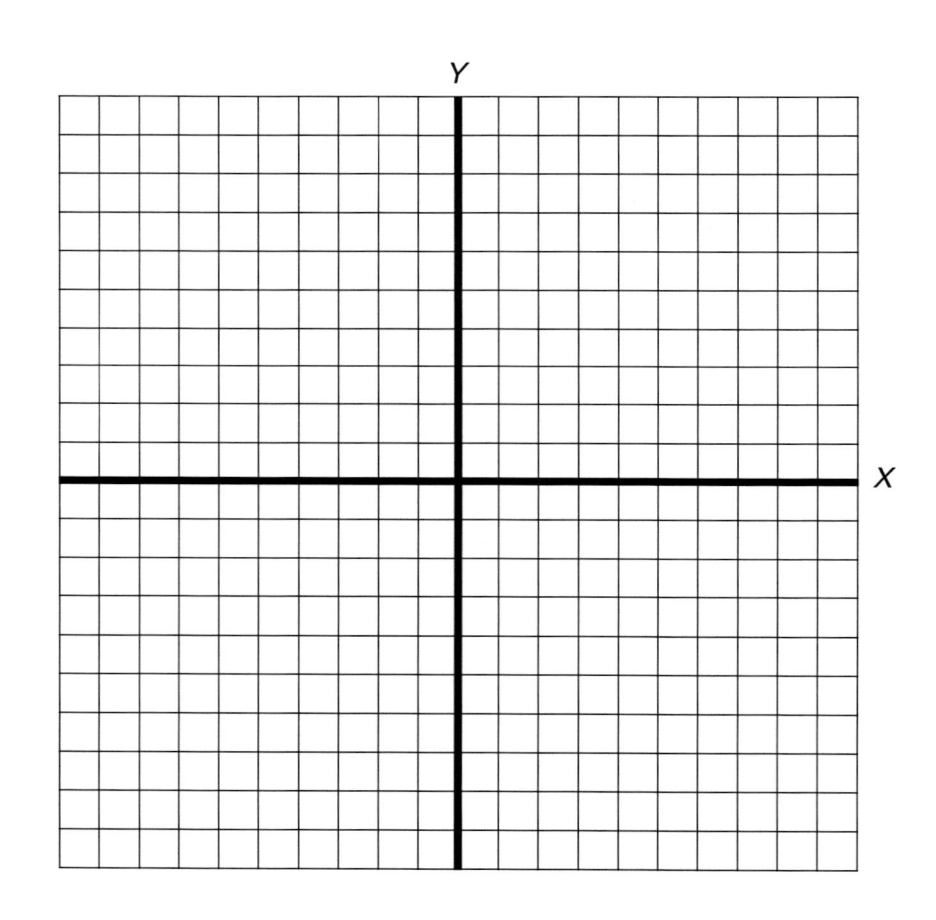

Name _____ Date _____

33. The Distance Formula

Find the distance between the following points:

1. (3, 4) and (0, 0) _____

2. (7, 6) and (5, 3) _____

3. (−4, 5) and (3, 2) _____

4. (7, 5) and (4, −3) _____

5. (5, 6) and (15, 21) _____

6. (−2, 3) and (3, −9) _____

7. (4, 4) and (7, −17) _____

8. (9, 0) and (9, 5) _____

9. (−4, −4) and (4, 4) _____

10. What expression could be used to represent the distance between the point (X, Y) and the point (3, 6)?
    _____

11. What expression could be used to represent the distance between the point (X, Y) and the point (P, Q)?
    _____

© 1996 J. Weston Walch, Publisher

Algebra Practice Exercises

Name _____   Date _____

34. Synthetic Division

Find the quotient and the remainder in the following division problems:

1. $\dfrac{X^3 + 3X^2 + 3X + 1}{X + 1}$ _____

2. $\dfrac{3X^3 + 2X^2 + 5X + 1}{X + 2}$ _____

3. $\dfrac{2T^2 + 3T + 1}{2T + 1}$ _____

4. $\dfrac{7Y^4 + 12Y^2 + 3}{Y^2 - 2}$ _____

5. $\dfrac{3X^4 - 5X^3 - 5X^2 + 4X - 1}{3X - 2}$ _____

6. $\dfrac{2X^4 - 3X^3 + 9X - 8}{X - 4}$ _____

7. $\dfrac{X^5 - 1}{X - 1}$ _____

8. $\dfrac{2X^6 + 3X^5 - 7X^3 + 3X - 1}{X - 3}$ _____

9. $\dfrac{6X^4 - 5X^2 - 13}{2X + 5}$ _____

10. $\dfrac{3X^5 - X^4 + X^3 - 6X^2 + 12X + 8}{X - 3}$ _____

11. $\dfrac{4X^5 + 7X^3 - 7X}{2X - 1}$ _____

12. $\dfrac{6X^5 - 12X^4 + 16X^3 + 32X^2 + 47X - 88}{2X + 3}$ _____

© 1996 J. Weston Walch, Publisher   Algebra Practice Exercises

35. Two Variable Equations and the Chart Graphing Method

Find three different points on the Cartesian plane that satisfy the following equations:

1. $3X - 3Y = 6$

2. $2Y + 7 = 3X$

3. $4Y - 5X = 2$

4. $2X - Y = 4$

5. $4X + 3 = 5Y$

6. $X = 8$

7. $6X + 11 = 5Y$

8. $10X - 3Y = 14$

9. $Y = -2$

10. $X + 3Y = 6$

Name _____ Date _____

36. The Slope of a Graph

The slope of a line is described by the measurement between two points in the following format:

Slope of line is equal to the ratio $\dfrac{\text{change in } Y\text{-direction}}{\text{change in } X\text{-direction}}$

For instance, in the lattice graph below, the slope of the line from the point *f* to the point *p* is −1/4, because to go from *f* to *p*, you must go down (negative) 1 unit and across (positive) 4 units.

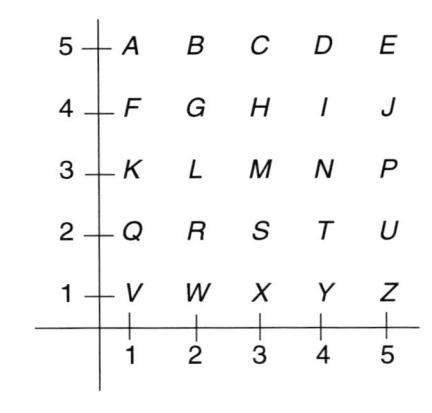

```
5 + A   B   C   D   E
4 + F   G   H   I   J
3 + K   L   M   N   P
2 + Q   R   S   T   U
1 + V   W   X   Y   Z
    +---+---+---+---+
    1   2   3   4   5
```

Find the following slopes:

1. The line from *L* to *J* _____
2. The line from *A* to *Z* _____
3. The line from *J* to *L* _____
4. The line from *M* to *Q* _____
5. The line from *F* to *H* _____
6. The line from *K* to *W* _____
7. The line from *V* to *U* _____
8. The line from *S* to *J* _____
9. The line from *A* to *E* _____
10. The line from *D* to *M* _____
11. The line from *K* to *I* _____
12. The line from *S* to *X* (Be careful!) _____

Name _____ Date _____

37. The Slope/Y-Intercept Method

Put the following equations into slope/Y-intercept form to determine the slope of their graphs:

1. $Y - 4 = 2X$ _____
2. $X + 2Y = 5$ _____
3. $3X - Y = 3$ _____
4. $2X = 5 - 4Y$ _____

Put the following equations into slope/Y-intercept form to determine their Y-intercepts:

5. $6 = 3X - 2Y$ _____
6. $5X = 12 - 4Y$ _____
7. $3X + 2Y = 5$ _____
8. $X - 4Y = 7$ _____

Sketch the graphs of the following equations by the slope/Y-intercept method:

9. $Y = 2X - 3$

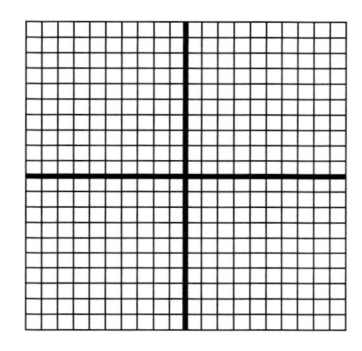

10. $3Y + 2X = -6$

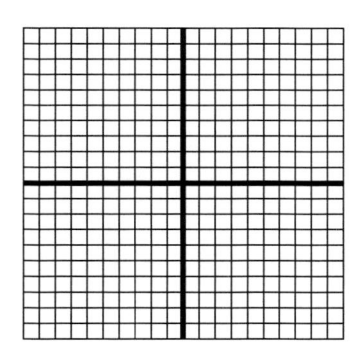

11. $4X - 3Y = 2$

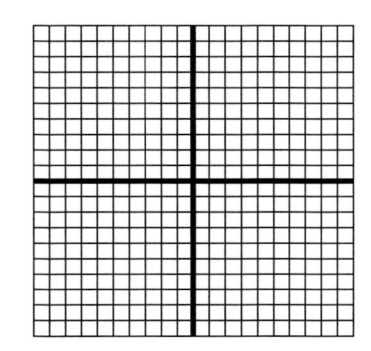

© 1996 J. Weston Walch, Publisher

Algebra Practice Exercises

Name _____ Date _____

38. Finding the Equation from the Graph

Find the equations of the graphs pictured below:

1.

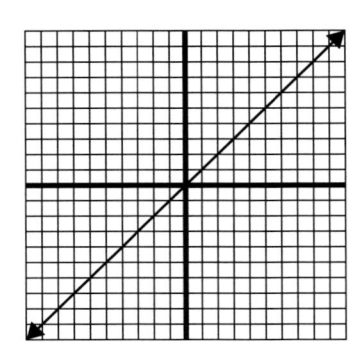

4.

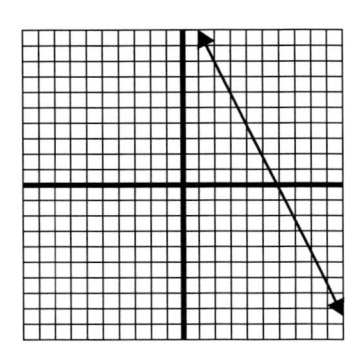

2.

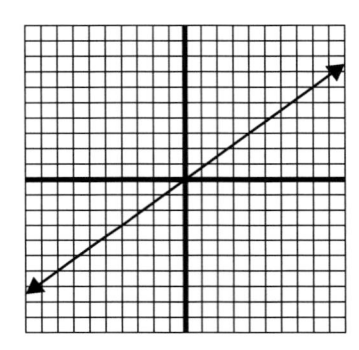

5.

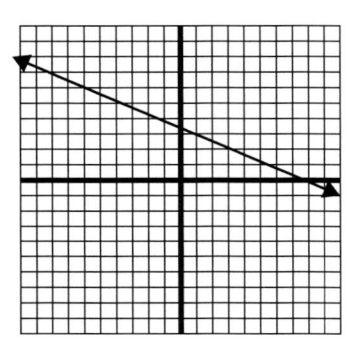

3.

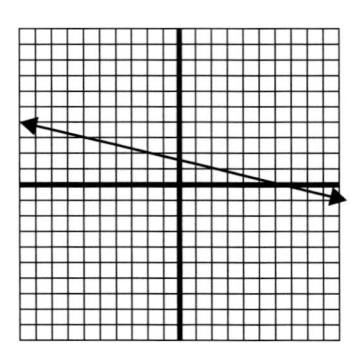

6.

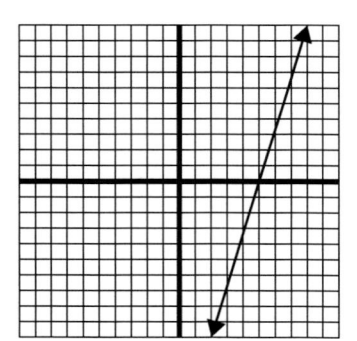

(continued)

© 1996 J. Weston Walch, Publisher   38   *Algebra Practice Exercises*

Name _____ Date _____

38. Finding the Equation from the Graph (continued)

7.

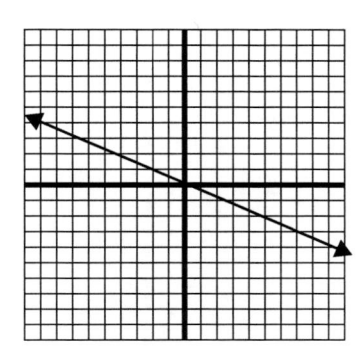

9.

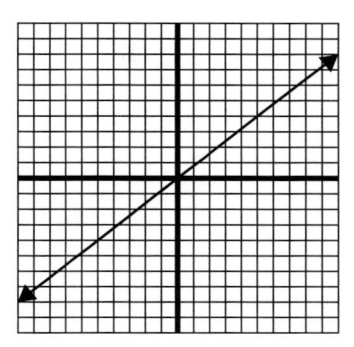

8.

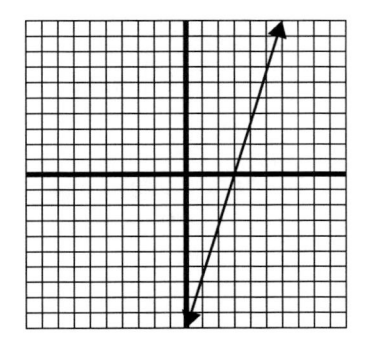

10.

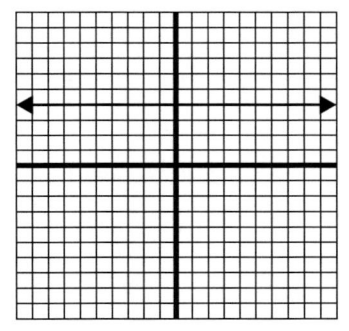

Name _____ Date _____

39. Parallels and Perpendiculars

Graph the following:

1. The line perpendicular to $4Y - 2X = 6$ that passes through $(3, 1)$

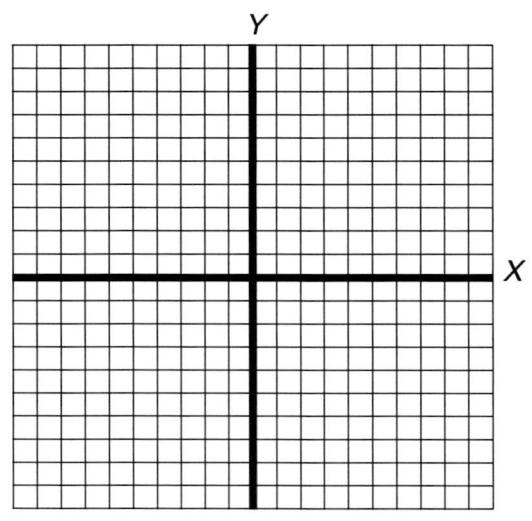

2. The line perpendicular to $3Y = X + 3$ that passes through $(2, 2)$

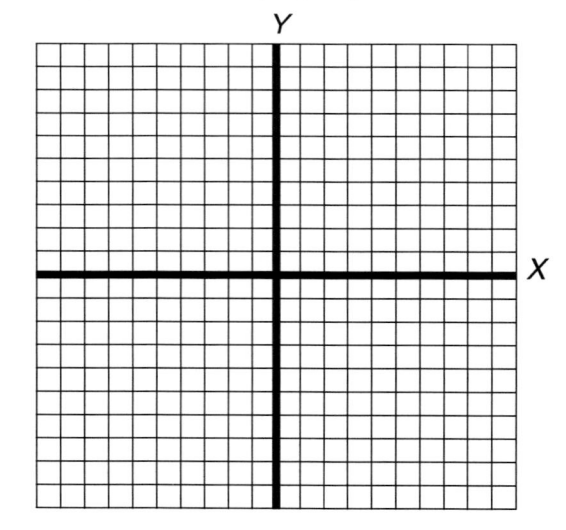

3. The line perpendicular to $7 - 3X = 5Y$ that passes through $(0, 0)$

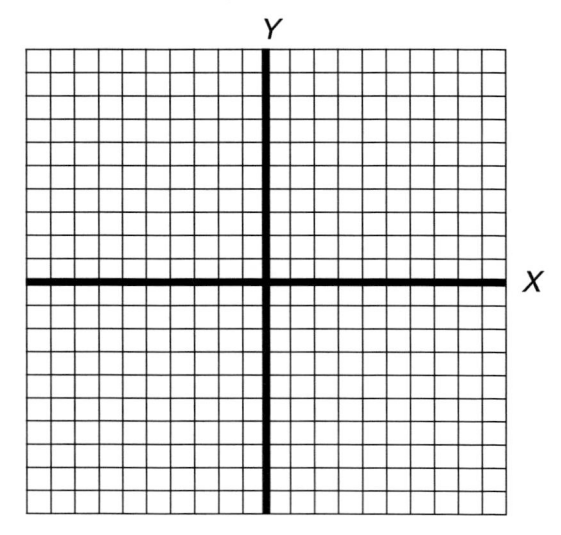

4. The line perpendicular to $4X = 3Y - 6$ that passes through $(4, 0)$

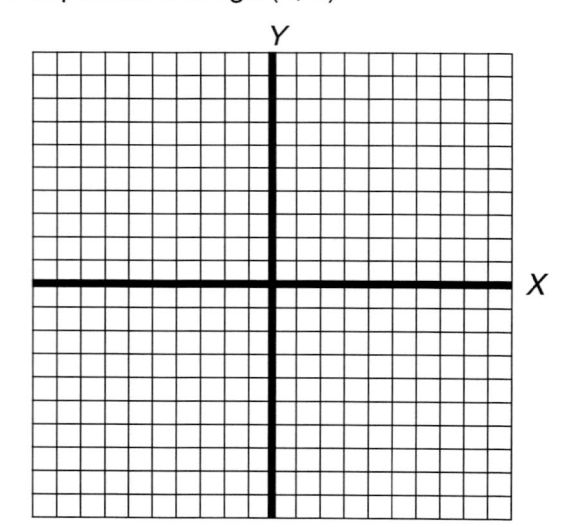

(continued)

Algebra Practice Exercises

Name _____ Date _____

39. Parallels and Perpendiculars (continued)

5. The line perpendicular to $Y + 3X = 9$ that passes through $(-1, -3)$

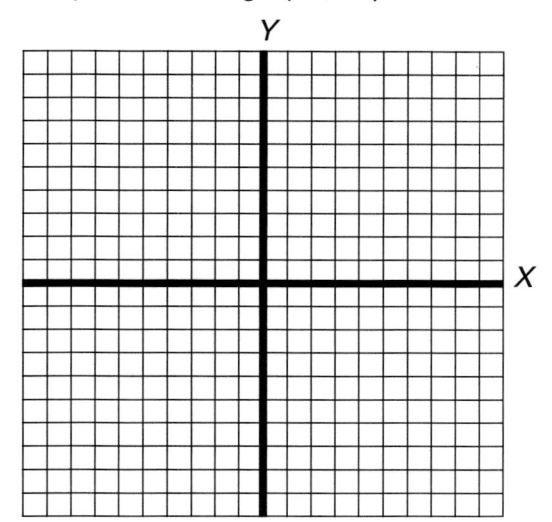

7. The line parallel to $Y - 4X = 8$ that passes through $(0, 2)$

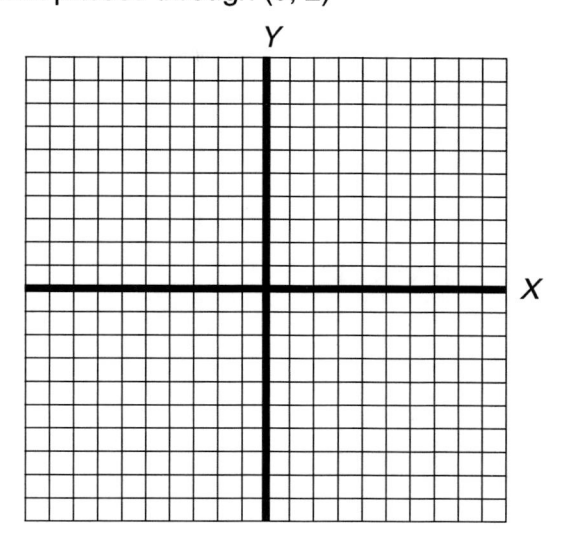

6. The line parallel to $2Y + 3X = 5$ that passes through $(-2, 5)$

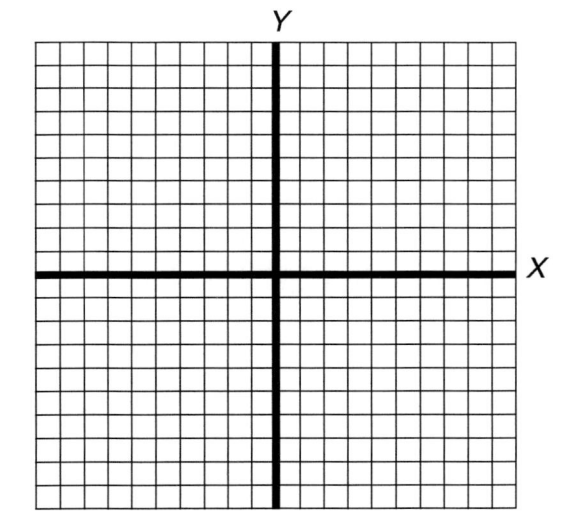

8. The line parallel to $2X - 5 = 2Y$ that passes through $(4, -1)$

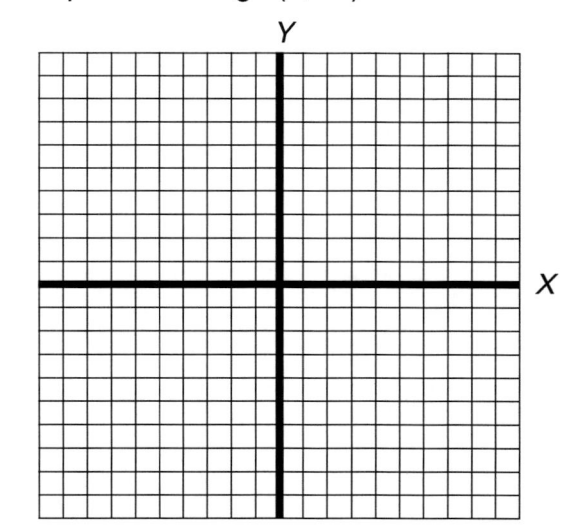

(continued)

© 1996 J. Weston Walch, Publisher   41   Algebra Practice Exercises

39. Parallels and Perpendiculars (continued)

9. The line parallel to $3X - 2Y = 4$ that passes through $(3, 2)$

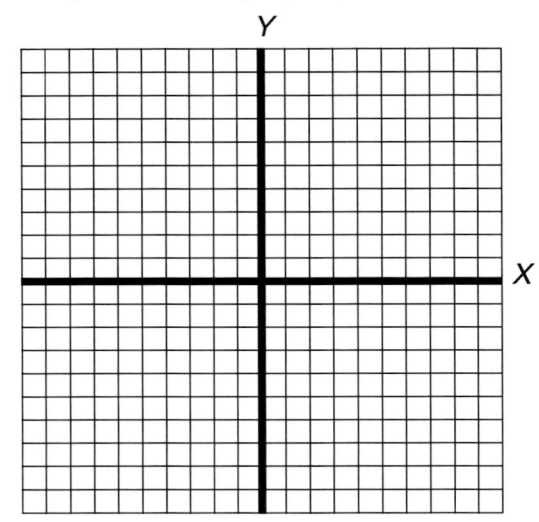

10. The line parallel to $4Y = 6X$ that passes through $(-2, 1)$

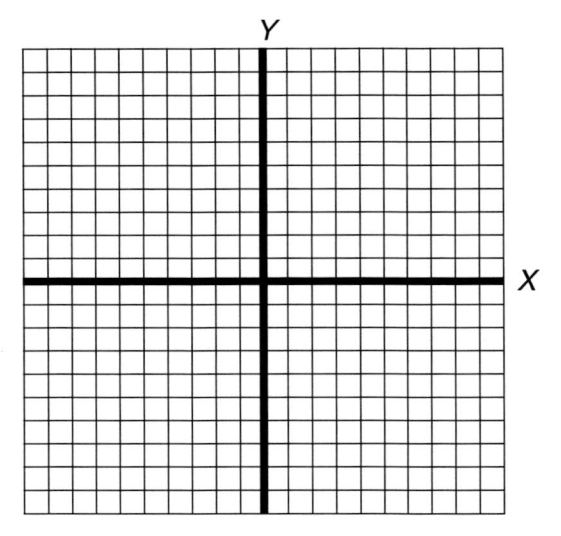

Name _____ Date _____

40. Finding Intersections by Plotting the Graphs

Plot the following sets of equations and estimate the intersections:

1. $2X - 3Y = 5$ and $3X - Y = 5$

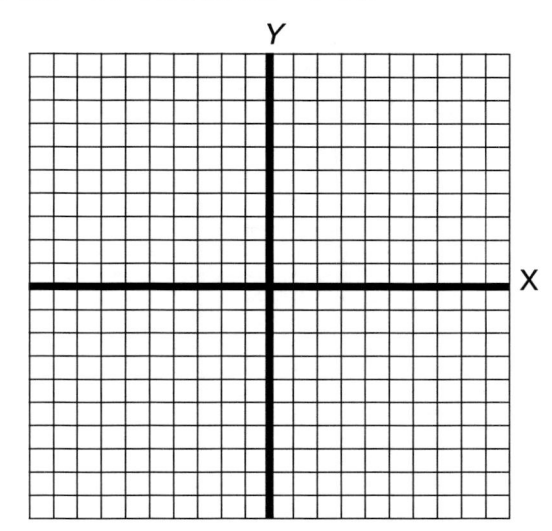

3. $4Y - 5 = 2X$ and $-5X = Y$

2. $4X - 7 = Y$ and $2X + Y = 15$

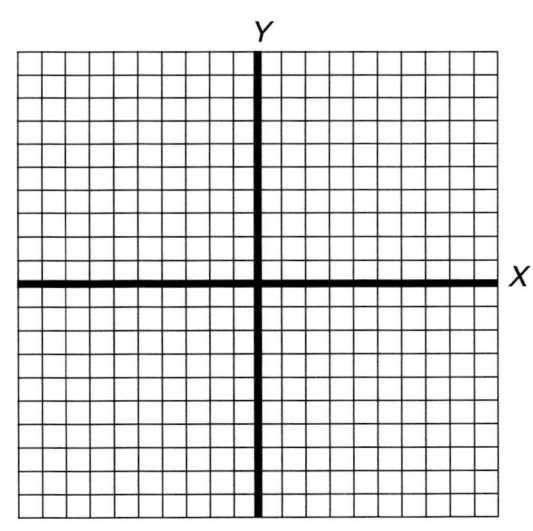

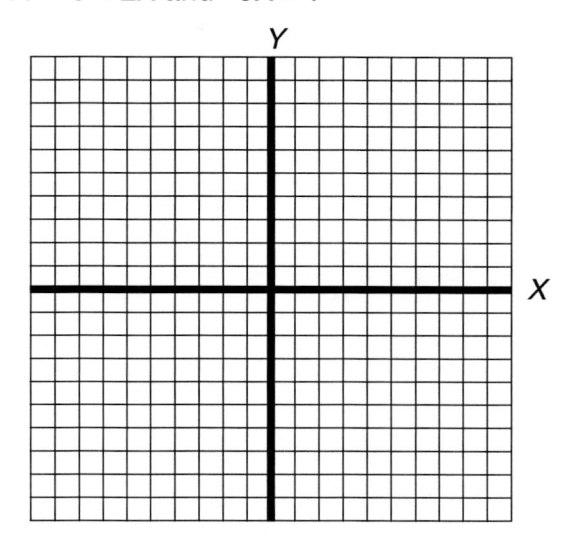

4. $5X - 2Y = 7$ and $3Y - 4X = -1$

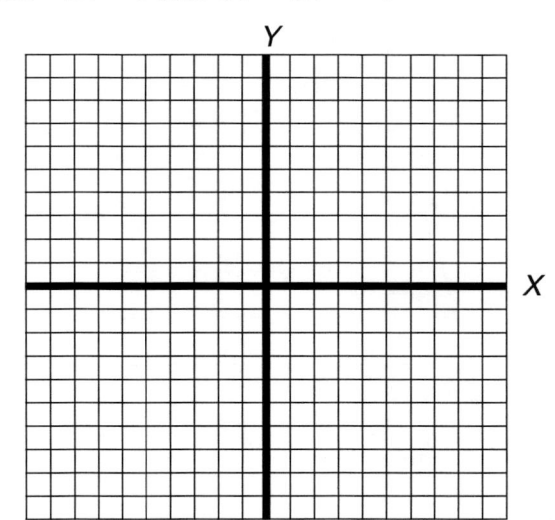

(continued)

© 1996 J. Weston Walch, Publisher

Algebra Practice Exercises

40. Finding Intersections by Plotting the Graphs *(continued)*

5. $X - 7Y = 0$ and $3Y = 2X$

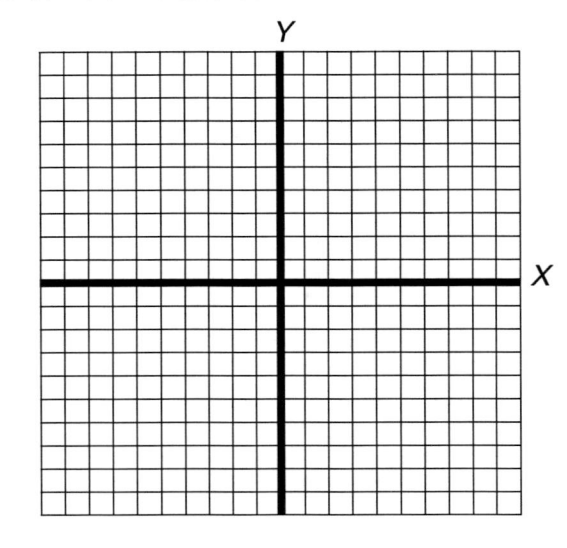

7. $2X - Y = 1$ and $3X - 2Y = 2$

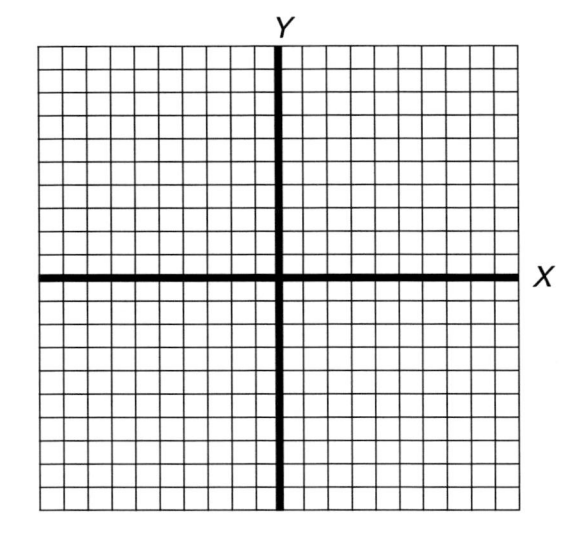

6. $X - 3Y = -9$ and $2X + 5Y = -3$

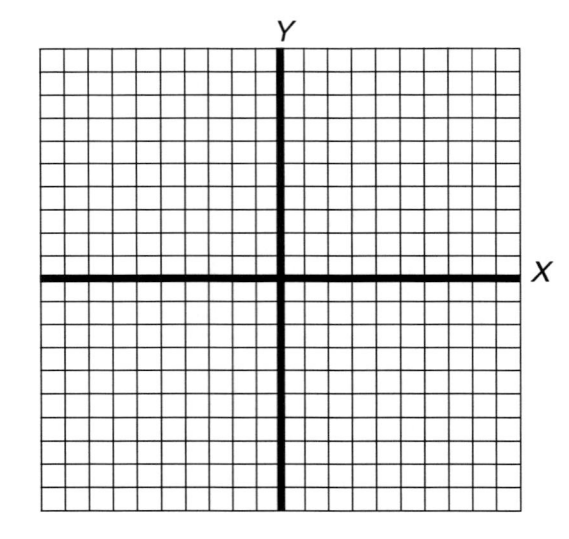

8. $Y + 3X = 13$ and $3Y - 3 = 2X$

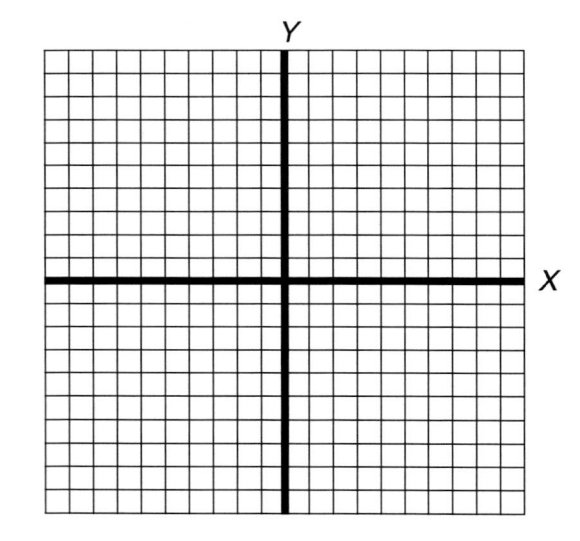

(continued)

Name _____ Date _____

40. Finding Intersections by Plotting the Graphs (continued)

9. $3X - Y = 2$ and $5X + 1 = 3Y$

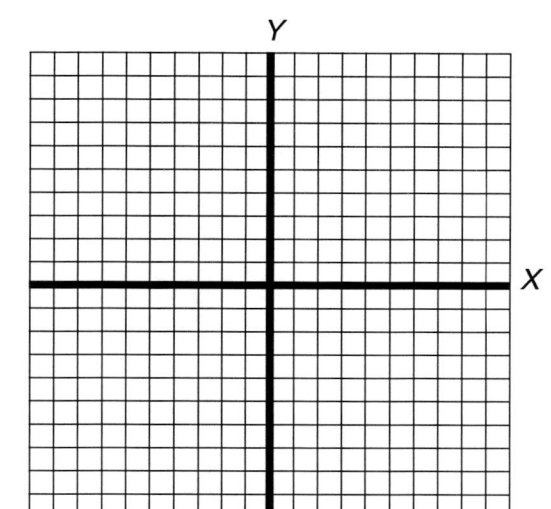

10. $6X - 5Y = 3$ and $3X - 1 = 14Y$

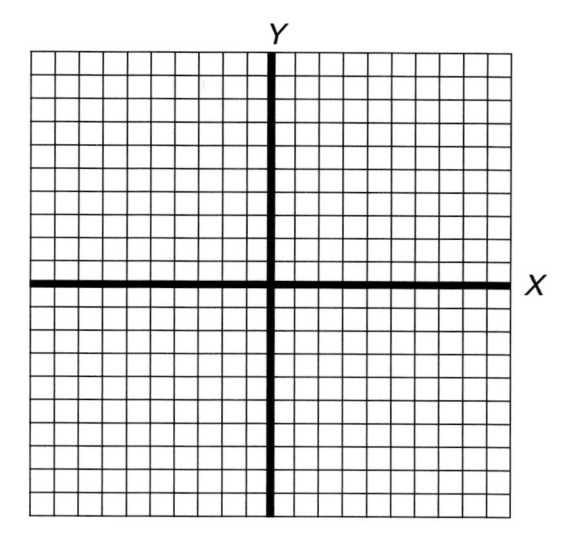

Name _____ Date _____

41. Solving Simultaneous Equations by the Substitution Method

Solve the following simultaneous equations by the substitution method:

1. $2X - 3Y = 5$ and $2X + 5Y = -3$ _____

2. $4X + 7 = Y$ and $3X - Y = 5$ _____

3. $4Y - 5 = 2X$ and $Y + 3X = 13.5$ _____

4. $5X - 2Y = 7$ and $X - 3Y = -9$ _____

5. $X - 7Y = 0$ and $3X - 1 = 14Y$ _____

6. $3X - 3 = 5Y$ and $2X + Y = 15$ _____

7. $.5X + 1 = 3Y$ and $3X - 2Y = 2$ _____

8. $2X - Y = 1$ and $3Y - 3 = 2X$ _____

9. $3X - 2Y = 2$ and $3Y = 2X$ _____

10. $6X - 5Y = 3$ and $3Y - 4X = -1$ _____

11. Find two equations that share (5, 3) as a solution. _____

12. Find two equations that share (–2, –3) as a solution. _____

© 1996 J. Weston Walch, Publisher

Algebra Practice Exercises

Name _____ Date _____

42. Solving Simultaneous Equations by the Elimination Method

Solve the following simultaneous equations by the elimination method:

1. $2X - 3Y = 5$ and $2X + 5Y = -3$ _____

2. $4X + 7 = Y$ and $3X - Y = 5$ _____

3. $4Y - 5 = 2X$ and $Y + 3X = 13.5$ _____

4. $5X - 2Y = 7$ and $X - 3Y = -9$ _____

5. $X - 7Y = 0$ and $3X - 1 = 14Y$ _____

6. $3X - 3 = 5Y$ and $2X + Y = 15$ _____

7. $.5X + 1 = 3Y$ and $3X - 2Y = 2$ _____

8. $2X - Y = 1$ and $3Y - 3 = 2X$ _____

9. $3X - 2Y = 2$ and $3Y = 2X$ _____

10. $6X - 5Y = 3$ and $3Y - 4X = -1$ _____

11. $3X - 4Y = 5$ and $2Y + X = 3$ _____

12. $2X + 3Y = 8$ and $3X - Y = 2$ _____

© 1996 J. Weston Walch, Publisher *Algebra Practice Exercises*

Name _____ Date _____

43. Solving Word Problems

Solve the following word problems:

1. Tyler is three times as old as Shawn and Ryan. In five years, Tyler will only be twice as old as Shawn and Ryan. How old are Shawn and Ryan? _____

2. Terri has 18 more marbles than Greg. After playing for a while, Terri has won 6 marbles from Greg and now has exactly twice as many as Greg has. With how many marbles did Terri start? _____

3. Ross is 32 years older than Ian. If Ross is twice as old now as Ian will be in 15 years, how old is Ross now? _____

4. Lori and Susan each build cabinets. Susan can make three times as many cabinets each week as Lori. If Lori works for two weeks, she can build 18 cabinets. How many cabinets does Susan build each week? _____

5. Michael and Adam have basketball card collections. Michael has 30 more cards than Adam. If Adam gave Michael 6 more cards from his collection, Adam would have half as many cards as Michael. How many cards does Adam have now? _____

6. Muriel and Richard love to play each other in Ping-Pong™. So far, Muriel has won four times as many games as Richard has. If they play 25 more games and she continues to win 80% of the games, she will have won 100 games. How many games have they already played? _____

7. Katie and Molly collect butterflies. Katie has five times as many butterflies as Molly, who just started her collection. If Katie gave Molly 16 of her butterflies, Katie would have three times as many butterflies as Molly. How many butterflies does Katie have right now? _____

8. Greg is four-and-one-half times as old as Margy. In two years, he will be three-and-one-third times as old as Margy. How old is Margy now? _____

9. Steve and Anne have a large collection of CD's. Steve bought three times as many of the CD's as Anne. If their collection numbers 236 right now, how many CD's did Steve buy? _____

10. Rory and Carol each bought a car. Rory's car had six times as many miles on it as Carol's car. They took a vacation together in Carol's car and drove 1,240 miles. Now Rory's car has five times as many miles as Carol's car. How many miles does Carol's car have on it now? _____

© 1996 J. Weston Walch, Publisher   Algebra Practice Exercises

44. Scientific Notation

Represent the following numbers in scientific notation:

1. 38.35

2. 47.8

3. 4,096

4. 16.00

5. 50,010

6. 0.0820

7. 1,048,576

8. 0.00007600003

Represent the following numbers in decimal form:

9. 3.4×10^{-3}

10. 7.8008×10^7

11. 6.701×10^{-5}

12. 2.35×10^8

13. 7.501×10^{12}

14. 3.14×10^{-11}

Simplify the following and give answers in scientific notation:

15. $\dfrac{(5.0 \times 10^{-3})(1.4 \times 10^8)}{2.0 \times 10^4}$

16. $\dfrac{(3.2 \times 10^{-7})(3.0 \times 10^{11})}{1.2 \times 10^{-5}}$

17. $\dfrac{(6.3 \times 10^5)(2.0 \times 10^{-4})}{1.8 \times 10^{-6}}$

18. $\dfrac{.00320 \times 450{,}000}{90{,}000}$

19. $\dfrac{72{,}000{,}000 \times 230}{18{,}400{,}000}$

20. $\dfrac{.00404 \times .0007}{.00000101}$

45. Simplifying Rational Expressions

Simplify the following:

1. $\dfrac{-7AB}{-21A^2B^2}$

2. $\dfrac{X^{t+2}}{X^t}$

3. $\dfrac{72J^2K^2}{-9JK^3}$

4. $\dfrac{(U^2V^3)^2}{5UV^5}$

5. $\dfrac{f^3+1}{e^2-1} \times \dfrac{e+1}{f+1}$

6. $\dfrac{r^2 \cdot r^n}{r \cdot r^{(n-1)}}$

7. $\dfrac{(m^2g^4)^3}{g^5m^4}$

8. $\dfrac{3Z^2}{2x^3} \cdot \dfrac{4x}{6Z^5}$

9. $\dfrac{n^{k-3}m^{3k}}{n^{k+2}(m^{k-1})^2}$

10. $\dfrac{k^4}{m^2} \cdot \dfrac{m^3}{b^7} \cdot \dfrac{b^5}{k^2}$

11. $\dfrac{2D^{-3}}{A^5b^{-2}}$

12. $\dfrac{2G^{-2}H}{G^2H^{-2}}$

13. $\dfrac{(wz)^{-3}}{z^2} \cdot \dfrac{1}{z^3w^{-5}}$

14. $\dfrac{m^2-9}{m^2+4m+3}$

15. $\dfrac{c^2-6c}{c^2+2c-48}$

16. $(t^4+5t^3+6t^2)(t^2+2t-3)^{-1}$

17. $\dfrac{r^4-s^4}{r^3-r^2s+rs^2-s^3}$

Name _____   Date _____

46. Solving Simple Inequalities

Solve the following inequalities on the number lines provided:

1. $3X - 2 < 2X + 5$ <--|--|--|--|--|--|--|--|--|--|--|--|-->

2. $2Z + 7 \leq 5Z - 11$ <--|--|--|--|--|--|--|--|--|--|--|--|-->

3. $4Y - 6 > 3Y - 1$ <--|--|--|--|--|--|--|--|--|--|--|--|-->

4. $X + 3 \leq 4X - 6$ <--|--|--|--|--|--|--|--|--|--|--|--|-->

5. $8N - 7 \geq 4N + 5$ <--|--|--|--|--|--|--|--|--|--|--|--|-->

6. $3(W - 4) \geq 2(W + 5) + 3$ <--|--|--|--|--|--|--|--|--|--|--|--|-->

7. $2(Z - 2) < \frac{2}{3} Z$ <--|--|--|--|--|--|--|--|--|--|--|--|-->

8. $4(2 - R) - 3(2R - 4) > 1$ <--|--|--|--|--|--|--|--|--|--|--|--|-->

9. $2(X - 3) - .5(X + 2) \leq 0$ <--|--|--|--|--|--|--|--|--|--|--|--|-->

10. $4(3X - 1) - 3(X + 2) < 7X + 6$ <--|--|--|--|--|--|--|--|--|--|--|--|-->

© 1996 J. Weston Walch, Publisher   Algebra Practice Exercises

Name _____   Date _____

47. Solving Compound Inequalities

Solve the following compound inequalities and graph your solution on the number lines provided:

1. $2Y + 3 < 5$ or $3Y + 1 \leq 3$ <--|--|--|--|--|--|--|--|--|--|--|--|-->

2. $-4 < 2D + 7 < 3$ <--|--|--|--|--|--|--|--|--|--|--|--|-->

3. $2K + 1 < 3$ or $3K - 6 > 6$ <--|--|--|--|--|--|--|--|--|--|--|--|-->

4. $-6 \leq 2V + 4 \leq 3$ <--|--|--|--|--|--|--|--|--|--|--|--|-->

5. $3Y < -6$ and $2Y - 1 > -19$ <--|--|--|--|--|--|--|--|--|--|--|--|-->

6. $-5 \leq 3T + 11 \leq 5$ <--|--|--|--|--|--|--|--|--|--|--|--|-->

7. $5B - 3 < 7$ and $2B + 12 > 2$ <--|--|--|--|--|--|--|--|--|--|--|--|-->

8. $Y - 4 \leq 7 + 2Y < Y + 17$ <--|--|--|--|--|--|--|--|--|--|--|--|-->

9. $4 + 7P \geq 9P - 3 \geq 7P - 6$ <--|--|--|--|--|--|--|--|--|--|--|--|-->

10. $2 - 6H \leq 4 - 8H \leq 4H - 6$ <--|--|--|--|--|--|--|--|--|--|--|--|-->

© 1996 J. Weston Walch, Publisher   Algebra Practice Exercises

Name _____ Date _____

48. Solving Compound Absolute Value Inequalities

Solve the following absolute value inequalities and graph your solution on the number lines provided:

1. $|4J - 3| < 5$ <--|--|--|--|--|--|--|--|--|--|--|--|--|-->

2. $|2R - 6| > 8$ <--|--|--|--|--|--|--|--|--|--|--|--|--|-->

3. $|X + 1| \geq 5$ <--|--|--|--|--|--|--|--|--|--|--|--|--|-->

4. $|2F - 2| < 5$ <--|--|--|--|--|--|--|--|--|--|--|--|--|-->

5. $3|T - 2| \leq 9$ <--|--|--|--|--|--|--|--|--|--|--|--|--|-->

6. $|5Y + 2| - 3 > -1$ <--|--|--|--|--|--|--|--|--|--|--|--|--|-->

7. $4|2P - 7| \leq 8$ <--|--|--|--|--|--|--|--|--|--|--|--|--|-->

8. $2|T - 8| + T > 7$ <--|--|--|--|--|--|--|--|--|--|--|--|--|-->

9. $3|Z + 5| - 3 < 6$ <--|--|--|--|--|--|--|--|--|--|--|--|--|-->

10. $|4A + 6| - 7 \leq 4$ <--|--|--|--|--|--|--|--|--|--|--|--|--|-->

© 1996 J. Weston Walch, Publisher   Algebra Practice Exercises

49. Graphing Linear Inequalities

Graph the following inequalities:

1. $Y > 3$

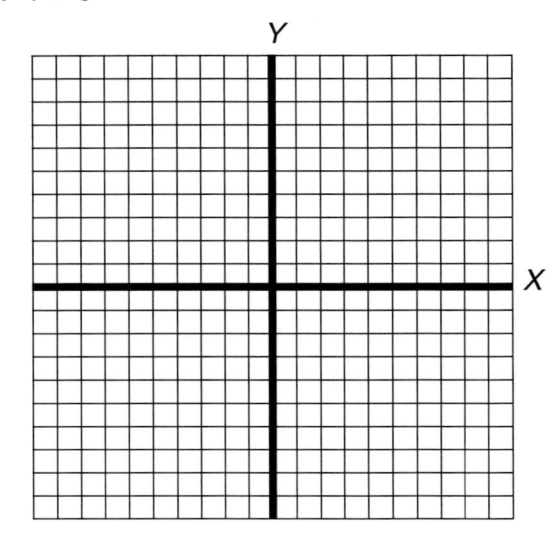

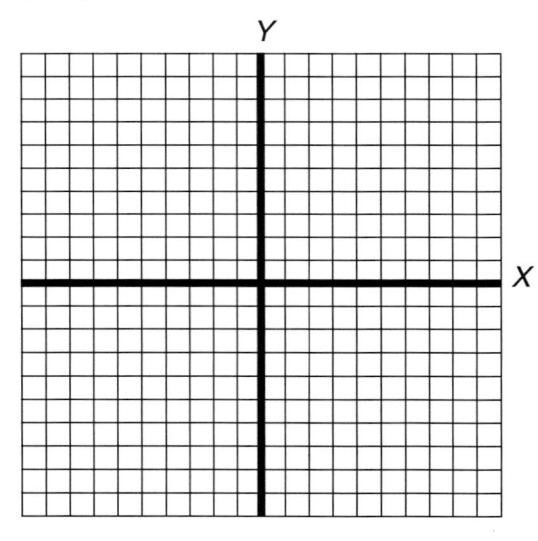

3. $X + Y \leq 5$

2. $X \geq -5$

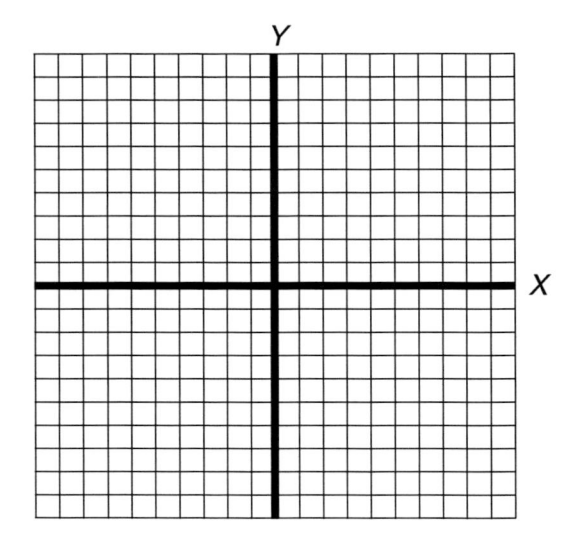

4. $Y > 6 - X$

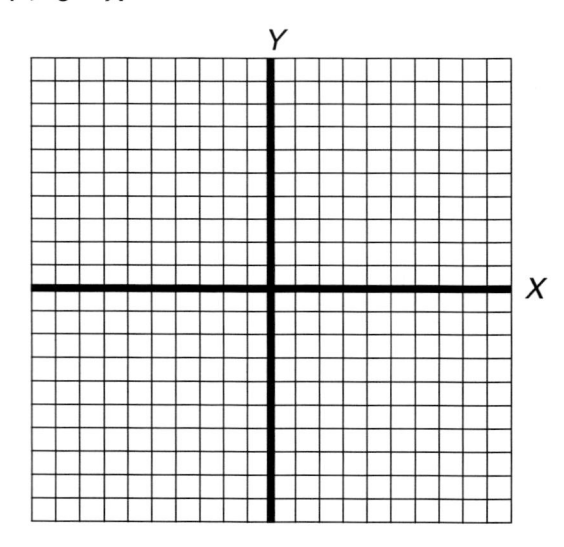

(continued)

Name _____ Date _____

49. Graphing Linear Inequalities (continued)

5. $3X - 4Y < 0$

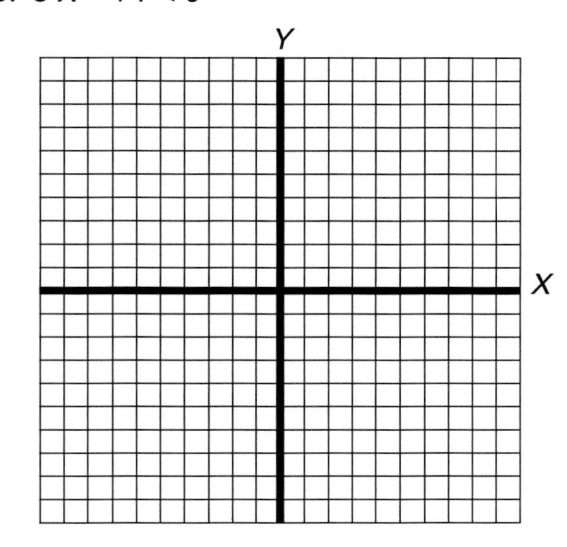

7. $2X - 3Y \geq 8$

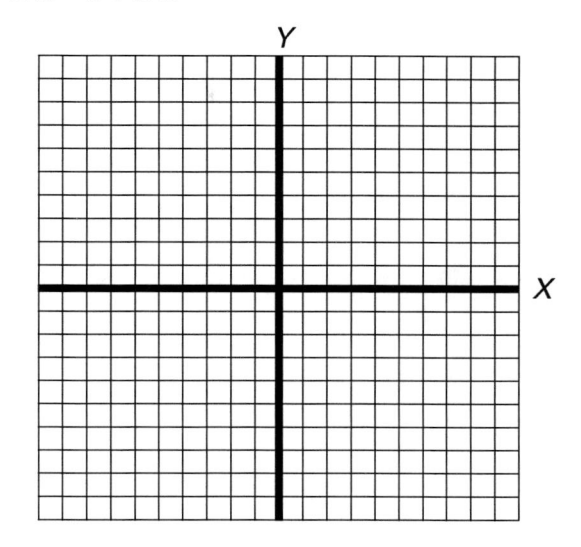

6. $3Y - X > -2$

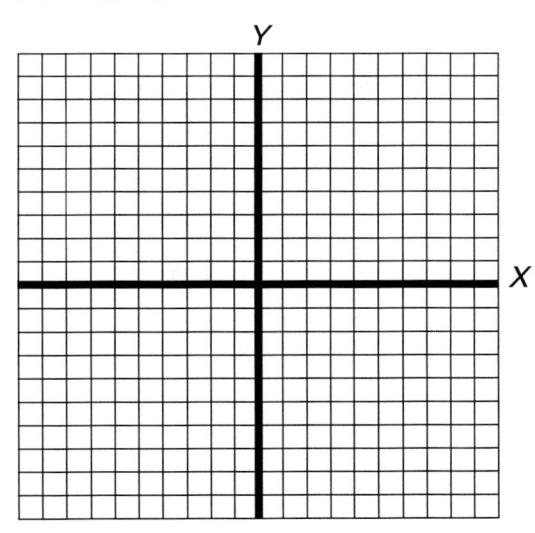

8. $2Y - 5 > 3X - 6$

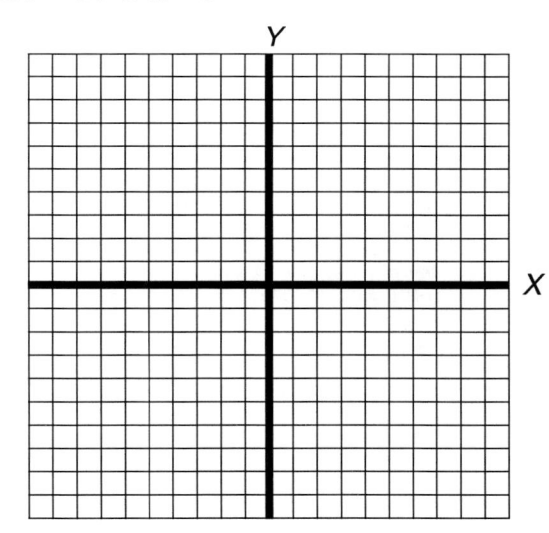

(continued)

Algebra Practice Exercises

Name _____ Date _____

49. Graphing Linear Inequalities (continued)

9. $4(X + Y) \geq 2X - 7$

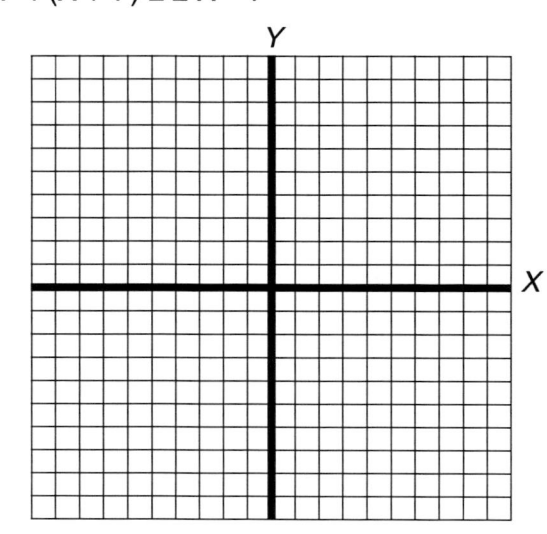

10. $3Y + 3 \leq 2(2Y - X + 3)$

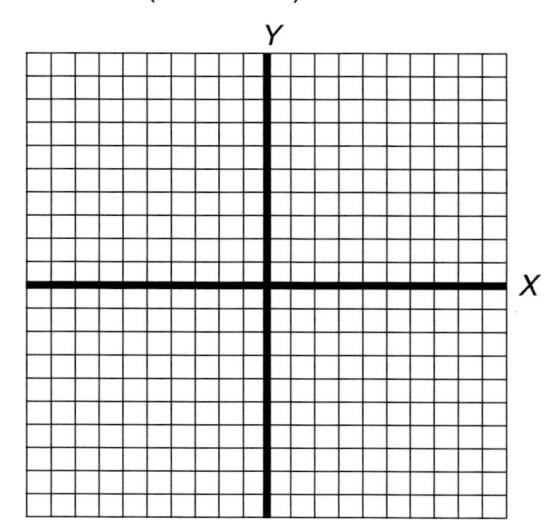

In the pictures below, identify the inequalities represented by the shaded region:

11.

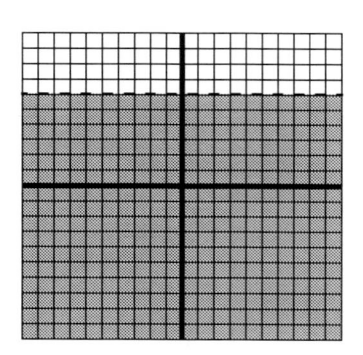

12.

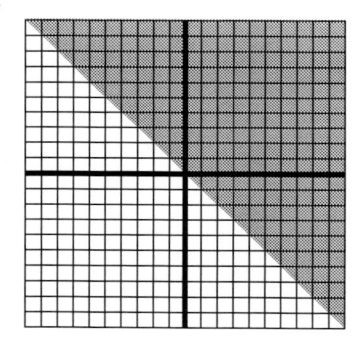

© 1996 J. Weston Walch, Publisher

50. Graphing Systems of Linear Inequalities

Graph the following sets of inequalities to find the region of intersection:

1. $Y \geq 4$ and $X < -1$

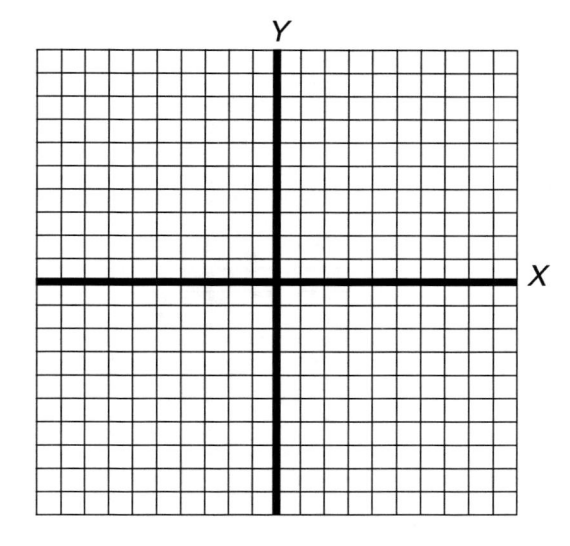

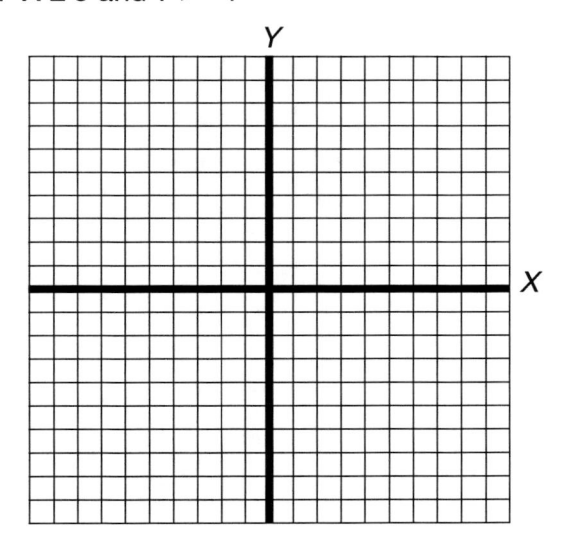

3. $X \leq Y - 4$ and $Y < 6$

2. $X \leq 5$ and $Y > -4$

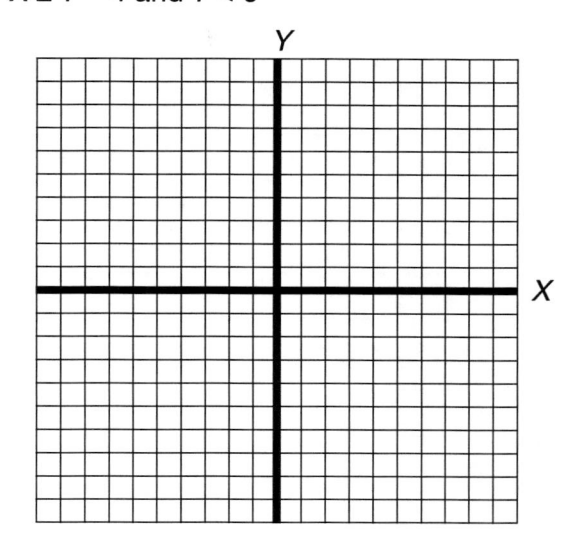

4. $2X - 3 > Y$ and $X > 1$

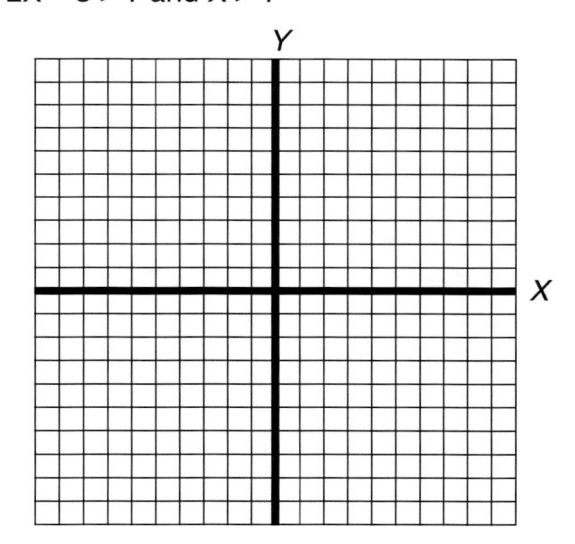

(continued)

Algebra Practice Exercises

50. Graphing Systems of Linear Inequalities (continued)

5. $X + 2Y < 12$ and $4X - Y < 4$

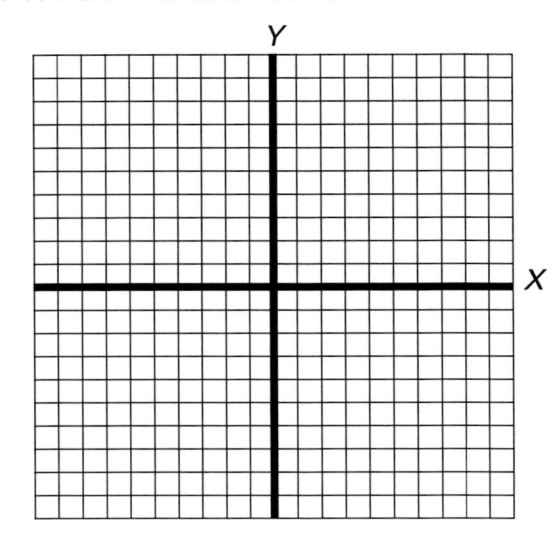

6. $X - 3Y < 6$ and $3X - Y > 2$

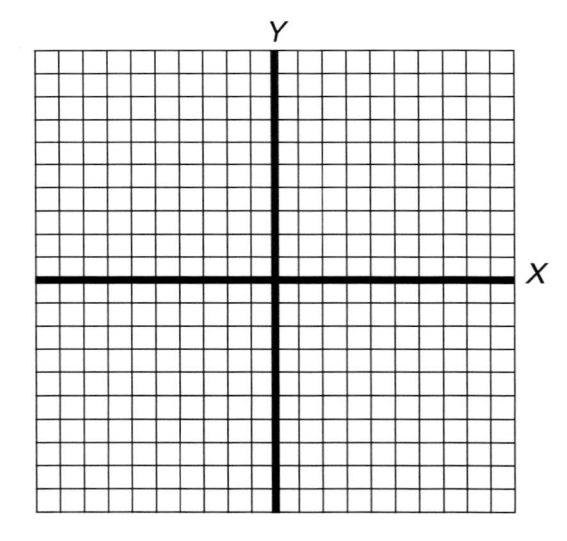

7. $2Y + 3X > -8$ and $X \geq 3 - 2Y$

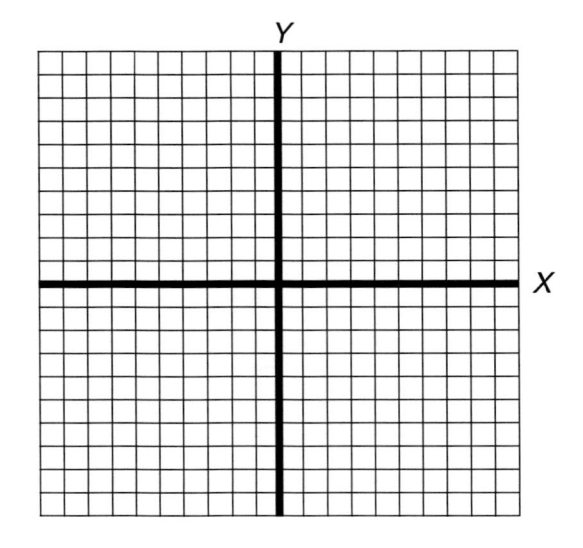

8. $X < 3Y - 2$ and $Y \geq 2X - 1$

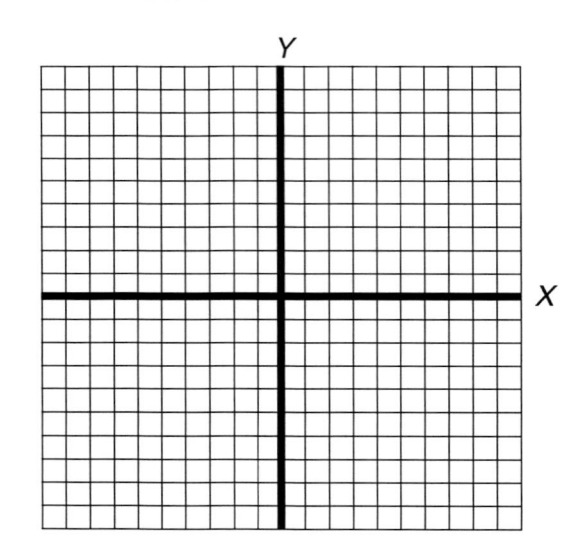

(continued)

Name _____   Date _____

50. Graphing Systems of Linear Inequalities (continued)

9. $2X - Y < 7$ and $4X - 3 > 2Y$
   (Careful!)

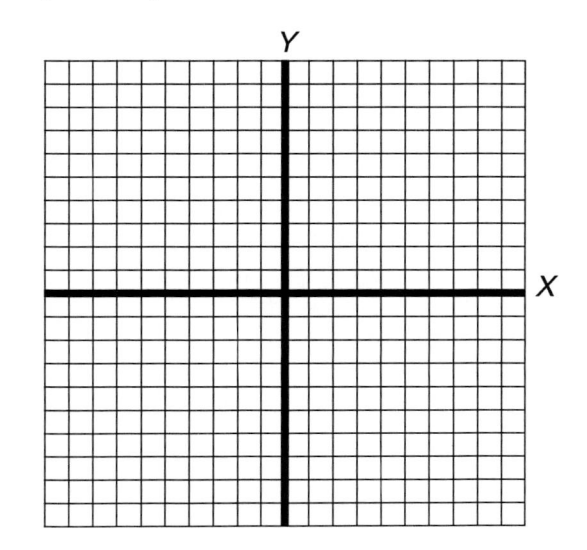

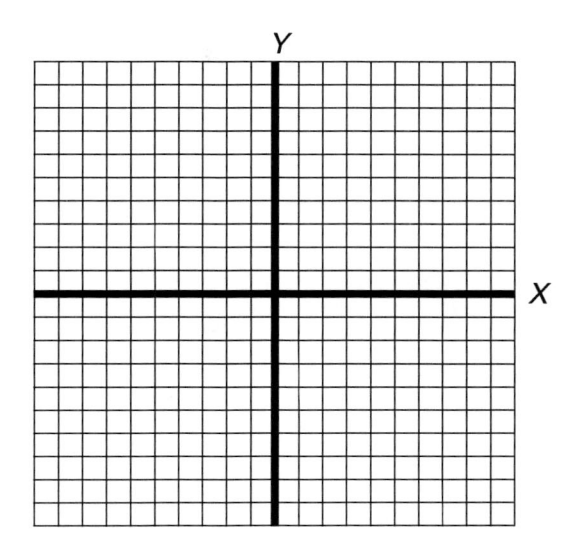

10. $3 - 5Y \leq 2X$ and $6X + 2Y \leq 9$